FOREST CERTIFICATION IN SUSTAINABLE DEVELOPMENT

Healing the Landscape

FOREST CERTIFICATION IN SUSTAINABLE DEVELOPMENT

Healing the Landscape

Chris Maser • Walter Smith

LEWIS PUBLISHERS

Boca Raton London New York Washington, D.C.

Library of Congress Cataloging-in-Publication Data

Maser, Chris.
    Forest certification in sustainable development : healing the landscape / by Chris Maser and Walter Smith.
        p.   cm.
    Includes bibliographical references.
    ISBN 1-56670-510-X (alk. paper)
    1. Forest management—Standards. 2. Forest products—Certification. 3. Sustainable forestry. I. Smith, Walter. II. Title.

SD387.S69 M37 2001
634.9′068—dc21
                                                   00-056300
                                                     CIP

© 2001 by CRC Press LLC
Lewis Publishers is an imprint of CRC Press LLC

No claim to original U.S. Government works
International Standard Book Number 1-56670-510-X
Library of Congress Card Number 00-056300
Printed in the United States of America 1 2 3 4 5 6 7 8 9 0
Printed on acid-free paper

Foreword

The concept of "certification" for forest management came about as a result of (1) concerns over the elimination of rain forests with no plan for replacement in kind, (2) concern for the net total impact of major disruption of large ecosystems and watersheds that can result from conversion of natural forests to agricultural or urban expansion, (3) concern for the social needs of indigenous peoples in developing countries, and (4) a recognition by concerned professionals of the benefits of wood as a renewable product of nature that can be processed with far less overall environmental impact than will result from making steel, plastic, aluminum, and cement.

These were the reasons for finding a means of ensuring that wood as a usable product for sustainable development could be encouraged. To do this, there needed to be some form of auditing harvesting practices and regeneration programs. An international group of people from multiple disciplines of business, the environment, and social services came together to address the interrelationship issues of forest utilization. Their goal was to develop a process that would allow some "endorsed" utilization of the products of forests by defining those principles and criteria that would minimize environmental impact while still achieving social stability.

The Forest Stewardship Council was the product of the work of these visionaries. In 1993, the Forest Stewardship Council became a functioning organization, and the ten principles and criteria for achieving endorsement for good forest management were promulgated.

The founders of the Forest Stewardship Council realized that it was necessary to recognize and address the reality of issues concerning the environment, society, and economics in order to gain a sustainable balance. Overemphasizing any one area could be detrimental to them all.

The concept of outsiders inspecting the activities of forestry professionals has been controversial in the economic and governmental sectors from day one. It was this concern that led the founders of the Forest Stewardship Council to develop a three-chambered organization — Environmental, Social, and Economic. It is also the reason that the certification process is divided into three overview categories — renewability, biological diversity, and socioeconomic benefits. Although this trilateral certification process is designed to achieve the best possible balance, it will never be perfect from the solitary perspectives of any of the three disciplines.

The Forest Stewardship Council is an international organization. For that reason, it is not a governmental process. Membership is strictly voluntary. Both the forest owner and the wood consumer can be the beneficiaries of forest management certification.

Collins Pine Company was the first corporation in the United States to engage in certification under the Forest Stewardship Council's principles and criteria and carries the differentiated products into the marketplace.

Consumers are consistently displaying their awareness of environmental impacts. Recycling has truly become a watchword for people in many countries. Education through quality information is becoming increasingly critical for consumer acceptance of products. It is virtually impossible for any consumer to reject the concepts embodied in the ten principles and criteria of the Forest Stewardship Council's certification program. The principles and criteria are designed to recognize and honor the rights of indigenous peoples, to recognize and obey the laws of each country, to require good planning, to consider the health and integrity of the ecosystem as a whole, and — in essence — to emulate nature.

In a relatively short span of time (10 years), the concepts of the Forest Stewardship Council have been implemented and forest management practices have been reviewed and endorsed in over 50 million acres of forestland in 40 countries on 5 continents. The United States represents nearly 6 million acres of this total in over 60 forest properties.

Incorporating more sensitivity toward the overall environment is gaining momentum around the world. It is therefore timely that Chris Maser and Walter Smith provide an extensive dialogue on the background of forestry practices in the United States that concern ecologists, environmentalists, and social servants. Although some people may think their critique to be harsh — particularly toward governmental and industrial forestland managers — foresters operating under Forest Stewardship Council certification are not likely to return to prior practices. In addition to their critique of forestry practices, Maser and Smith also provide a very detailed explanation of the principles and criteria of the Forest Stewardship Council under the SmartWood program.

Embarking on a journey of sustainability can begin with forest certification. To many, the journey will look like the unknown faced by Columbus' crew in 1492. Is there an edge to the world? Taking the plunge into forest certification is, for some reason, of great concern to foresters and forest owners. Will the owner and the forester lose control of the property? This seems to be the principle restraint keeping many from embarking on the journey. The answer is on the horizon. I hope Maser and Smith will help chart the course.

James E. Quinn
President and CEO
The Collins Companies

The Authors

Chris Maser spent over 20 years as a research scientist in natural history and ecology in forest, shrub steppe, subarctic, desert, and coastal settings. Trained primarily as a vertebrate zoologist, he was a research mammalogist in Nubia, Egypt (1963–1964) with the Yale University Peabody Museum Prehistoric Expedition and was a research mammalogist in Nepal (1966–1967) for the U.S. Naval Medical Research Unit #3 based in Cairo, Egypt, where he participated in a study of tick-borne diseases. He conducted a 3-year (1970–1973) ecological survey of the Oregon coast for the University of Puget Sound, Tacoma, Washington. He was a research ecologist with the U.S. Department of the Interior, Bureau of Land Management, for 12 years (1975–1987), the last 8 years studying old-growth forests in western Oregon, and a landscape ecologist with the Environmental Protection Agency for a year (1990–1991).

Today he is an independent author as well as an international lecturer and a facilitator in resolving environmental disputes, vision statements, and sustainable community development. He is also an international consultant in forest ecology and sustainable forestry practices.

He has written over 260 publications, including the following books: *Forest Primeval: The Natural History of an Ancient Forest* (1989, listed in the *School Library Journal* as best science and technical book of 1989), *Global Imperative: Harmonizing Culture and Nature* (1992), *Sustainable Forestry: Philosophy, Science, and Economics* (1994), *From the Forest to the Sea: The Ecology of Wood in Streams, Rivers, Estuaries, and Oceans* (1994, with James R. Sedell), *Resolving Environmental Conflict: Towards Sustainable Community Development* (1996), *Sustainable Community Development: Principles and Concepts* (1997), *Mammals of the Pacific Northwest: From the Coast to the High Cascades* (1998), *Setting the Stage for Sustainability: A Citizen's Handbook* (1998, with Russ Beaton and Kevin

Smith), *Vision and Leadership in Sustainable Development* (1999), *Reuniting Economy and Ecology in Sustainable Development* (1999, with Russ Beaton), and *Land-Use Planning for Sustainable Development* (2000, with Jane Silberstein). Although he has worked in Canada, Egypt, France, Germany, Japan, Malaysia, Nepal, Slovakia, and Switzerland, he calls Corvallis, Oregon home.

Walter Smith is not only an ex-logger and past owner of a logging company but also a professional certification and training consultant who provides technical expertise and chain-of-custody certification assessment and audits on forest management.

His love of all things forestry has taken him from surveys as a forestry assistant in the late 1960s and the manufacture of logs in a lumber mill, to the training of professional timber fellers at the college level and for the U.S. government. These early training experiences and his unique interest in forest practice compliance led eventually to his consultancy in forestry for government and forestry organizations, which have included the Sierra Club, the California Department of Forestry, and the Institute for Sustainable Forestry. As this organization's Certification Program Coordinator, he coauthored the forestry and harvesting standards and guidelines for implementation, and he led teams of professionals in assessing forest management for certification. He has conducted certification assessor training and conferences in the U.S., Canada, and other countries including Indonesia, Malaysia, Japan, and China, and he has extensive experience in certification assessment and reporting on forestland management in industry.

He is a founding board member of the Redwood Forest Foundation, Inc., a nonprofit organization dedicated to buying and owning forestlands for the benefit of the local community. He served as interim director of the Foundation from 1998 to 1999 and, in that capacity, was their public spokesperson. Mr. Smith is an advisory board member of the Pacific Forest Trust, which endeavors to protect working forestland through the establishment of conservation easements. He is also a founding member of the Forest Stewardship Council (FSC), an international standards-setting body that accredits certifiers to assess excellence in forestry worldwide.

Acknowledgments

It is with pleasure that we thank Steve Radosevich (Professor of Forestry Science, Oregon State University, Corvallis), James E. Quinn (President and CEO of The Collins Companies, including Collins Pine in Portland, Oregon), and Richard Z. Donovan (Director, SmartWood Program, Rainforest Alliance, Richmond, Vermont) for reviewing our manuscript and making numerous improvements.

We extend a special "thank you" to our wives, Zane and Karen, for their patience while we were engrossed in writing this book, and to Zane for proofreading the final copy.

Dedication

To the memory of Fred Furst, my friend and neighbor when I was a boy in the 1940s. Fred, who retired as the Supervisor of the Siuslaw National Forest in western Oregon, had been in the U.S. Forest Service from the end of World War I into the mid-1940s. And it was Fred, regaling me with stories of the early days as I followed him around his huge garden, who taught me to love the forest and the whole notion of the U.S. Forest Service and forestry.

— Chris

To my family: my father and mother, Ed and Ruth Smith; my wife Karen; and my daughters Summer, Fauna, and Trinity who taught me everything I know and who had enough patience and love to stick with me through some interesting career changes. Second, to loggers who work so hard and get such a bad rap, yet are the essence of forest management and the foundation of forest-dependent communities. Last, to the staff at the Institute for Sustainable Forestry and SmartWood, both of whom opened their hearts and minds to me, an old logger, and allowed me to pursue challenges that have gone beyond what I could have dreamed of.

— Walter

If you are a poet, you will see clearly that there is a cloud floating in this sheet of paper. Without a cloud, there will be no rain; without rain, the trees cannot grow; and without trees, we cannot make paper. The cloud is essential for the paper to exist.

— Thich Nhat Hanh, Vietnamese monk

Contents

Introduction

Certification as an assurance of product quality to the consumers by a disinterested third party assessor is a concept that has been around for some time. Underwriters Laboratory is perhaps the most recognizable certification label, which is affixed to electrical appliances that meet or exceed a certain standard of quality and safety. The Forest Stewardship Council is similar in that its label of certification notifies the consumer that certain ecological, social, and economic standards of biological and social sustainability have been met by the forest landowner or forest management company using standards derived by a diverse group of professionals within the greater public arena for the larger public good.

Certification by the Forest Stewardship Council is totally voluntary on the part of the evaluee and the assessment is done by a certifier who is accredited through the Forest Stewardship Council. The voluntary nature of the forest certification is the strength of the program because feelings and values can neither be legislated nor regulated through legal means. Human feelings and values can, however, be taught and nurtured in a way that creates an intrinsic desire to do voluntarily what one feels to be morally correct within one's own conscience for oneself and future generations.

To better understand forest certification, we will discuss what forest certification is and how it got started.

What Is Forest Certification?

Like all things historical, we recount the inception of forest certification as we understand it, because all we can do is view the history of past events through our own lenses, which means our own interpretation of what we understand to have taken place.

Forest certification is a multifaceted program that includes standards for forest practices and management, marketing of forest products, and public education, all of which are subjected to a verifiable assessment by a third party to ensure that a given forest is being managed in as sustainable a manner as possible. First, forest certification is based on standards for forest practices and management that include three major components:

1. Standards for overall ecosystem management, including timber harvest, that describe such things as how many trees of what species and age-classes in a given distribution must be left alive and growing when harvest operations are over, the density and configuration of the forest road system, and the measures required to protect wildlife habitat, both terrestrial and aquatic, in order to maintain biological, genetic, and functional diversity

2. A systematic and documentable procedure, starting with a required long-term management plan that is peer reviewed by resource professionals and followed by several on-the-ground inspections to ensure that the forest owner has complied with the standards to which he or she has agreed

3. A "chain of custody" program for tracking products that have been certified as coming from a sustainably managed forest as they change hands through sequence of subsequent ownerships that go from the landowner → the sawmill → wholesaler of the lumber → the retailer → the consumer

Second, certification is a program for marketing forest-derived products. Forestlands that are being managed in accordance with the standards of biological sustainability are publicly recognized as such through the documented certification labels that are affixed to the products, such as logs, as they leave the forest. A unique market advantage, such as a premium price, has been created for producers and manufacturers of forest products that carry the certification label. In turn, a consumer-driven market can then provide an incentive for forestland owners to manage their lands in ways that will benefit their local communities economically (potentially, a premium price for logs to the local landowners and a greater market share for the local mill operator) and environmentally (a landscape that has greater ecological integrity and thus better protects the environmental wealth of the community, such as clean water and biological, genetic, and functional diversity within the overall landscape).

Third, a certification program is a safe, voluntary vehicle for public education. Both producers and consumers of forest products need the best available information regarding the sustainability of our forests and the importance of an alternative to the destructive management practices of forestry.

In short, forest certification is a way for society to articulate a working vision of both ecologically sound forestry and sustainable community development. It is thus important that the certifier faithfully apply the accepted standards of behavior to each forestry operation using community, professional, and academic assessors of forest resources and practices to determine whether the operations meet the standards. The assessor, in turn, must be willing to accept and adhere to the philosophical underpinnings of the standards. The goal is threefold: (1) develop a consensus among interested parties of what constitutes ecologically sound forestry; (2) identify those forestry operations that closely follow the established standards; and (3) promote those operations in the marketplace in order to reward them for their commitment to meeting or exceeding the standards.

Forest certification is not, however, meant as the end-all of conservation tools. There are other tools that can also help to promote ecologically sound forestry, such as conservation easements; protests; litigation; land acquisition; regulation; recycling all wood products, not just paper; remodeling a house, rather than tearing it down to build a new one; and building modest houses, rather than the gargantuan ones that are currently being constructed. Nevertheless, in deference to other avenues of change within forestry, we think forest certification has the capacity to be the most openly educational, emotionally friendly, socially gentle, and economically flexible of all the conservation tools.

Be that as it may, there are three problems that need to be transcended before certification can go "full steam ahead": (1) getting acceptance from the timber industry, which still sees certification as a plot to stop logging; (2) educating the consumer about the relationship of where the wood they purchase comes from and how it is grown and harvested and the social pressures on the forest; and (3) having a consistent supply of certified wood to enter the marketplace so retailers can offer it on a regular basis, i.e., having a supply of all of the grades, sizes, species, and products so that an entire project can be made using certified wood.

The Certification Movement

Over the last couple of decades, people in the U.S. have become used to seeing, and now often look for, the "ecolabels" or "green labels" on recycled paper products, but they seldom look for such labels on less processed products from the forest, such as lumber; this lack of awareness, however, is changing.

According to Eric Hansen, a specialist in marketing forest products with Oregon State University Extension Service, it is possible to find some European retailers who carry two-by-fours sporting a label from the Forest Stewardship Council, an international organization based in Mexico that oversees certification of forest practices and products. These labels are similar to those guaranteeing that vegetables in the marketplace have been grown organically. The label tells the consumer that the lumber has been certified by an independent third party to come from a forest, the care or "management" of which is based on sound ecological, social, and economic principles and practices.*

Market forces are converging and pushing the management of forests toward certifiably sound social–ecological principles in a social effort to bring about biologically sustainable forestry, which is the unequivocal basis of an economically viable forest industry that in turn nurtures the source of its viability — a biologically sustainable forest. (*Although totally true biological sustainability in forestry is probably not possible, excellence in the practice of forestry is, and it is precisely such excellence that certification is designed to help owners of forestland achieve!*) Hansen thinks the globalization of the economy, the strength of the "green movement" in Europe, and the creation of demand for certified lumber through buyers' groups may influence the forest products industry in the U.S. to ultimately move toward the acceptance of certification, something they are now largely resisting.

Although the certification of forest practices through the certification of forest products is in its infancy in the U.S., it is well established and growing ever stronger in Western Europe. In fact, Sweden may well have led the way.**

Sweden, as a nation, began as early as 1903 to project into the future a forest's ability to grow wood on a sustainable basis. Cutting young forests to accommodate the activities of mining companies was banned in 1923, and public pressure during the early 1970s led to a ban on the use of herbicides as a practice in forestry. A new forest management act, passed in 1993, placed equal emphasis on managing the forests of Sweden for wood *and* other values. "Sweden's work toward more environmentally friendly forestry," says Hansen, "called for restoring damaged land and setting aside some forested land for wildlife habitat." How did this all come about?

---

* OSU News Service. 1998. 'Ecolabels' could catch on with U.S. forest products. *Corvallis Gazette-Times*, Corvallis, OR. October 5.

** Eric Hansen, Rick Fletcher, and James McAlexander. 1998. Sustainable forestry, Swedish style, for Europe's greening market. *Journal of Forestry* 96(3):38–43.

Swedish foresters in the 1960s (like many foresters today) tried ignoring the growing criticism environmental groups leveled at them over their ecologically insensitive forestry practices by dismissing environmental groups as extremists not to be taken seriously. In the 1970s, when forestry practices were questioned in the media, foresters stood on their credentials saying they were the professionals and thus they alone understood forestry; they had everything under control, even herbicides. Despite such efforts, they lost the use of herbicides due to intolerance by the public. After two decades of conflict, Swedish foresters began in the 1980s to listen, learn, and cooperate with the environmentalists, who not only refused to go away but also held sway over public opinion. The cooperation of the late 1980s led to national collaboration in the 1990s in which a coalition of the timber industry, associations of woodland owners, indigenous peoples, labor unions, and groups of environmentalists joined together and formed the Swedish Forest Stewardship Council Working Group to collectively deal with forest management and planning in the field.

In Germany, publishers are not just interested in using recycled paper; they also want to see that the forests, where the trees are grown to make the pulp for the paper, are managed in an ecologically sound and environmentally friendly way. Meanwhile, buyer groups, which began in the U.K. and the Netherlands (not individual consumers), are driving the demand for certified "green" products in many parts of Europe.

And, finally, in the U.S., Home Depot President and CEO Arthur M. Blank announced in August 1999 that "Our company sells less than 10 percent of the lumber in the world, but it still is the largest single retailer of lumber in the world. Home Depot will use the power of its purchasing dollars to vote for products that do the most to preserve environmentally sensitive areas. We are asking our vendors to help us by dramatically increasing the supply of certified forest products."*

Ragnar Friberg, chief forester of Stora Forest and Timber in Sweden, has a lesson that foresters in general and the forest industry worldwide would do well to heed in thought, word, and deed: "Commitment to sustainable forestry must be real," Friberg says. "A media campaign to change attitudes will not work. They [the public] will find you out."

Third-party certification of forests and forest products in the U.S. has become a credible force in the forest industry. There are several reasons for this growth in forest certification: (1) the political climate is changing as people increasingly move into urban areas and bring their votes with them; (2) those same people are becoming more discerning when it comes

---

* Michael Cronk. 1999. Home Depot plans shift to certified lumber only. *Knight Ridder Newspapers*, In: *Corvallis Gazette-Times*, Corvallis, OR. August 30.

to what they purchase, and wood from certified forests will become evermore in demand; (3) new products are being invented, such as boards that are part wood and part plastic, that may well reduce the market demand for lumber in the U.S.; (4) both three and four above will change the current economic paradigm to one more in keeping with social–environmental sustainability; and (5) as the above-mentioned things change, small woodland owners will find more certainty in a paradigm of biological sustainability through forest certification than they did when forestry was controlled in large part by a few multinational timber corporations.

Certifying forestry practices and products as ecologically sound and people friendly works when it is an independent third-party process, precisely because the process is voluntary from beginning to end. In addition, on the market side, certification requires secure control of the products and informs the consumer about available choices. The voluntary nature of certification and its ability to educate are imperative not only to its long-term success but also to its immediate credibility. We say this because no person or body politic can legislate feelings or a sense of value; those must come from within each person. Feelings and the expression of the values they engender are passed from one human being to another by word and deed, not by dictation or coercion. Conversely, those who conduct the third-party certification process must be as open minded, personally gentle, educationally explicit, and as noncompetitive with other accredited certifiers as humanly possible.

The import on which the above paragraph is written was brought home to me (Chris) years ago while speaking to 60 or so loggers in the late 1970s about changing logging practices to be more sustainable. When I was finished an older logger came up to me and paused as he gathered himself.

"Sonny," he said, his bright blue eyes snapping under his thinning silvery hair, "I'm sure you have a good point there, but I just can't find it. Mostly," he continued with a broad grin, "I think you're just full of shit."

Taken aback, I asked him why he thought that.

"Well," he replied, "the way I think of a forest, you just don't make no sense nohow." With that, he winked at me and left.

His position describes the industrial paradigm pretty well. Without a dramatic shift in the philosophical underpinnings of our belief systems, new data simply have nowhere to fit. This said, we have struggled to present the problem of nonsustainable forestry practices, considered by many to be "mainstream forestry," without assigning blame. If we must be disparaging, then we as a society are to "blame" because, for the most part, we did not know that we were doing anything wrong.

Then, when we discovered the errors of our ways and tried to correct them, we got into a fight with one another about who was right, who had the authority to change direction and practices, and who had the power to change them. This fight proved three important points: (1) you cannot change a problem using the same level of consciousness that created the problem in the first place, e.g., authority (top down), power (money), and misplaced values (I'm right, you are wrong); (2) before practices will change, the philosophy behind them must change; and (3) a shift in philosophy cannot be coerced by any amount of authority. It is thus critical that we teach through the certification process instead of depending on restrictive regulations to change human behavior. While the former maintains human dignity, the latter is usually perceived as a form of punishment to be staunchly resisted.

We, Chris and Walter, embrace SmartWood as a certification process (although other programs are equally accredited by the Forest Stewardship Council) because the above-mentioned elements, including teaching, are not only personally important to us but also well encompassed therein.

Although we may repeat the following concept in one way or another, it is critical at this juncture to understand what is meant by "certification." Certification means that the owners and/or managers of forestlands have voluntarily met the standards of excellence in the practice of forestry, standards that have been arrived at through consensus by foresters, biologists, conservationists, interested citizens, indigenous peoples, and economists. These standards of excellence are based on the best available science and the highest biological, social, and economic principles. In addition, certification makes it possible to consciously forge a new path toward sustainability, whatever that may prove to be over the long run. Certification also gives us reference points with which to make course corrections along the way.

With the above in mind, it is our intent in this book to make people aware of the possibilities that exist and to further the cause of social–environmental sustainability now and in the future. If, however, you are looking for a strictly linear approach to the subject, you will not find it because systems are cyclical, and to understand them you must think like them.

We have chosen the process of a SmartWood affiliate, the Institute for Sustainable Forestry based in Willits, CA, as our explanatory model. We have done this because I (Chris) have spent many years studying ecosystems and have learned that most people understand principles and concepts best, whether ecological or social, when they are applied, through examples, to a real landscape, such as that of northwestern California and southwestern Oregon, as opposed to a generalized hypothetical one. Once the principles and concepts are understood, however, it becomes clear that they can be adjusted and made applicable anywhere.

As you read this book, bear in mind that we are telling it *as we see it*. We can do nothing else because neither we nor anyone else *knows* how it is. We will use personal experiences to illustrate various points, and we will identify the voice simply as "Chris" or "Walter."

We offer this book as an example of what can be done to heal the forests of the world when people seriously want to do so. Granted, we are writing about an ideal. In so doing, we must assume all people have sufficient consciousness to understand the necessity of healthy landscapes to nurture them with the wealth of Nature's renewable energy and products. Without a diversity of raw materials to convert into economic energy, there can be no sustainable community. Beyond that, without the free services provided by Nature in the form of pollination, clean water, clean air, sequestration of carbon, and fertile soils, which we take for granted and in so doing often ignore, the quality of human life in existing communities will gradually decline with each successive generation until it sinks into mere survival.

Forest certification can go a long way in helping to heal the landscape and sowing hope in the souls of labor-weary adults and children for their future as they understand it. Healing the landscape is a journey that can begin with the voluntary efforts of a single person working to heal even a single acre.

Although many people with a strong environmental focus would like to see a "quick fix" to heal the landscape, there is in Nature no such thing. Change, especially social change, takes time and progresses through many little extraordinary steps taken by ordinary people whose dignity is based on doing what they feel is right for all generations.

For our part, we feel that forest certification is the "right" way to proceed because it is built on discipline, dignity, and trust, all of which can and will change attitudes, perceptions, and behavior over time. This, in turn, will raise the collective level of consciousness to mend a problem created by less enlightened thinking. The journey has already begun in several countries besides the U.S., but it needs everyone's help because time is of the essence.

We extend a personal invitation to you to join the adventure. After all, the principles and concepts presented in this book can be applied to grasslands, deserts, arctic tundra, or wherever healing is needed. We have intentionally called the journey an adventure because we cannot help to heal any part of Nature without also healing ourselves and our community in like measure. And it is this measure that brings the whole of the world one step closer to being sustainable for all its inhabitants.

Sustainability, as we use it, is a way of thinking in which choices and some things of value from which to choose are passed from one generation to the next — from adults to children — as each generation grows and reproduces. In this sense, sustainability is an ideal that inspires individuals to act in an "other-centered" way, which becomes a matter of service to the future. The struggle toward sustainability must therefore be voluntary. We say this because thinking cannot be legislated or regulated through the coercion of laws. Sustainability requires the integrity of intuition, flexibility of rational logic, and the passion of commitment to the long term — things that laws often inhibit instead of encourage.

We have divided the book into two parts, because when something is broken it can only be fixed by elevating our thinking to a level of consciousness higher than that which originally caused the problem. The first part of this book is therefore a brief discussion of the problem: forestry as it is practiced today. We offer this overview simply as a point of departure for the second part of the book, which is a discussion of the higher level of consciousness (sustainable forestry through the process of third-party certification) required to correct the social–environmental problems associated with today's forestry.

Forestry as an Evolution in Consciousness

Thinking of forestry as an evolution in consciousness, Frederick J. Deneke, assistant director of the USDA Forest Service Cooperative Forestry Program in Washington, D.C., wrote in the January 1998 issue of the *Journal of Forestry* that he had of late "been stepping back and observing the drama being played out over the perceived appearances of good and evil in the practice of forestry in the United States."[1] The practice of forestry, as viewed by Deneke, is not a matter of good vs. evil, but rather a matter of human consciousness, which has been continually evolving when it comes to understanding our relationship with Nature. There is, says Deneke, a range of consciousness in a society, such as ours, at any given time that could be characterized as "the unenlightened, the mainstream, the leading edge, and voices crying in the wilderness."

The timber barons who leveled the great forests of the eastern half of the U.S. at the turn of the century would today be characterized, according to Deneke, as the unenlightened, whereas landowners engaged in a certain level of trusteeship of the land would probably represent the center of consciousness. On the leading edge were people like Gifford Pinchot, the first chief of the newly created U.S. Forest Service, while John Muir and Henry David Thoreau were the voices crying in the wilderness. The leading edge from the 1930s to the 1950s would have been represented by Aldo Leopold, Supervisor of the Carson National Forest in New Mexico and later professor of wildlife management at the University of Wisconsin in Madison, or perhaps by an advocate for wilderness like Robert Marshall.

The leading edge of Pinchot's time, contends Deneke, is no longer the leading edge, but rather "smack dab" in the *center* of today's forestry due largely to what is being taught in university schools of forestry.[2] According to a recent study, for example, professional foresters, such as those in the U.S. Forest Service, embrace the utilitarian land ethic of Pinchot's time more than do wildlife biologists and other specialists in natural resources,[3] although perhaps not to such an extreme as that expressed by Karl F. Wenger, President of the Society of American Foresters. Wenger wrote a commentary in the January 1998 issue of the *Journal of Forestry* on his perception of why forests need to be managed:

> The fact is, Nature knows nothing. Nature is deaf, dumb, blind, and unconscious. ... It reacts blindly and unconsciously according to the properties and characteristics of its components. These have no intrinsic values, since only the human race can assign values. Nature doesn't care what we do to it. ... Clearly, the people's needs are satisfied much more abundantly by managed than by unmanaged forests.[4]

In contrast to this narrow view, Deneke thinks the concepts of ecosystem management and sustainable development are further out on the leading edge of consciousness, where people are working together at the level of water catchments and landscapes to resolve issues concerning natural resources.

In contrast to the general thinking of today's foresters and the current leading edge, "a present-day example of a voice crying in the wilderness could be Chris Maser and his work with sustainable forestry." Somewhere in time, says Deneke, Maser's work to maintain the biological health of forests for all generations while simultaneously using them for human benefit may become mainstream, and new voices will emerge to cry in the wilderness. The challenge for the generations of the future will be to find and follow those voices based on a sound ethical foundation of human values informed by the latest scientific understanding.

"Blessed are those on the leading edge and the voices in the wilderness, who compassionately show us the way by their example," all the while knowing that the cause of the problems — and thus the answers — lie within us and will be derived from how we carry out our work and how we see and treat ourselves and one another, which is but a mirror reflection of our own world view. They teach us, says Deneke, that the solutions to environmental problems are not in projecting blame, but rather in working together to help one another maintain dignity as we clean up our own individual and collective backyards.

"Those on the leading edge are the early implementers of the ideas emanating from the voices in the wilderness." They integrate their minds and hearts as they work together — and occasionally alone — for practical and peaceful solutions for the common good. I (Chris) once had the privilege of meeting such a man on Vancouver Island, British Columbia, Canada.

One evening, well over a decade ago, I sat next to an older gentleman, in the true sense of the word, as I ate supper with a group of small woodland owners. My task over the next couple of days was to conduct a workshop on forest ecology for the woodland owners. Although we spent the first day indoors viewing slides and discussing how forests function, the second day was spent in the field on this particular gentleman's tree farm.

Because there were so many woodland owners present, they were divided into two groups, one of which I was with in the morning and the other group in the afternoon. It seemed we discussed almost everything imaginable as we wandered about the tree farm. The owner, on the other hand, was busy with his own part of the workshop's agenda, which was to give the participants a guided tour of his tree farm. Finally, at day's end, as we all gathered to go to supper, he came up to me.

"Can I visit with you for a moment?" he asked.

"Of course," I said.

"Well, what do you think of my place here? You know, I've made my living off of this forest for 50 years, and a good living it's been too. What advice can you give me to improve what I do? I know I can always improve."

I regarded him for a moment to see how seriously he wanted to know what I thought because some questions are really idle chatter with no response desired. Sensing that he was in earnest, I said, "Your tree farm is beautiful and very much like a manicured park. Having said this, however, I have no advice to give you, only some ideas to share, if you so desire."

"Yes, by all means."

For the next half hour or so, we wandered about the tree farm as I pointed out that there were no diseased or dying trees, no standing dead trees or snags, and no fallen trees rotting on the forest floor. In addition, the tree stumps that remained after cutting had gotten considerably smaller than those of the original old-growth trees. He said little at first, but then came a torrent of questions about just how he could incorporate coarse woody debris into his management plans and field operations as a reinvestment of biological capital into the soil for the sake of the soil's long-term health and fertility. His enthusiasm was almost overwhelming.

It was many years before I saw him again, but over those intervening years, I heard how he was telling everyone and anyone who would listen about the ecological importance of large stumps, large snags, and fallen trees in maintaining the health of the forest soil as well as providing habitat and nutrient cycling. He has over the years educated a great many people on his tree farm by actively accepting the role of a leading-edge owner of a small woodland. How many people whom he has taught will become leading-edge owners of woodlands? It is people like this gentleman who help to elevate the consciousness of society.

"What is going on at any point in time is not a matter of right versus wrong or good versus evil," says Deneke, but rather "a reflection of where the mainstream of consciousness is at that time." For a discussion about elevating the consciousness of mainstream forestry, see *Reinventing the Forest Industry* by Jean Mater.[5]

PART ONE: A BRIEF LOOK AT TODAY'S FORESTRY

I see in the near future a crisis approaching that unnerves me and causes me to tremble for the safety of my country. ... Corporations have been enthroned, an era of corruption will follow, and the money power of the country will endeavor to prolong its reign by working upon the prejudices of the people until the wealth is aggregated in a few hands and the Republic is destroyed.

— Abraham Lincoln

Chapter 1

The Philosophical Foundation of Today's Forestry: An Overview

We cannot have an economically sustainable yield of any forest product, such as wood fiber, water, soil fertility, wildlife, or genetic diversity, until we first have an ecologically sustainable forest, one in which the biological divestments, investments, and reinvestments are balanced in such a way that the forest is self-maintaining in perpetuity. Sustainability is thus additive. We must have a sustainable forest to have a sustainable yield; we must have a sustainable yield to have a sustainable industry; we must have a sustainable industry to have a sustainable economy; and we must have a sustainable economy to have a sustainable society. It all begins with a solid foundation — in this case, a healthy, biologically sustainable forest.

Today's forest practices are largely counter to sustainable forestry, however, because instead of training foresters to grow and nurture forests, we train tree farm managers to manage the short-rotation, economic tree farms with which we are replacing our indigenous forests. Forests have evolved through the cumulative addition of species diversity, which in turn creates structural diversity, which in turn creates functional diversity — all of which add up to the diversity, complexity, and stability of ecological processes through time. We are reversing the rich building process of that diversity, complexity, and stability by replacing forests with tree farms designed only with narrow, short-term, economic considerations. Every acre on which a forest is replaced with a tree farm is an acre that is purposely stripped of its biological, genetic, and functional diversity, its ecological sustainability, and is reduced to the lowest common denominator — simplistic economics.

The concept of a tree farm, a strictly economic concept, has nothing whatsoever to do with the biological sustainability of a forest. Under this concept, indigenous forests are replaced with tree farms of genetically manipulated trees accompanied by the corporate–political–academic promise that such tree farms are better, healthier, more viable, and more productive of wood fiber than are the indigenous forests, which evolved with the land over millennia. But "sustainable" means producing industrio-economic outputs as the forest gives us the biological capability to do so in perpetuity. In turn, this necessitates balancing withdrawals of products with bio-economic reinvestments of biological capital in the health of the forest, especially in the soil. It means maximizing the health of the forest and harvesting all products and enjoying all amenities thereof with humility and gratitude.

To accomplish biological sustainability, we must shift our historical paradigm from that of the exploitive, colonial mentality — "use it until it collapses, then someone else can deal with it" — to the paradigm of trusteeship. Much as we might wish otherwise, humanity, even Western industrialized society, is not in control of Nature. So if society is to survive as we know it, we must become trustees of our natural resources — in the original sense of the word "resource."

Re and *source* mean reciprocal, to use something from the Earth and then to be the source of its renewal. Today's dictionaries define "resource" as any property that can be converted into money. Yet if we go back to the original sense of the word "re-source," we will find that the biological sustainability of our forests lies embodied in a word that we blithely use but do not fully understand.

Defining the Problem

Before discussing forest certification, it is necessary to expose the problems with the philosophical underpinnings of forestry that brought about the perceived need for forest certification in the first place. In laying out the problems, we (Chris and Walter) are fully aware that not all timber companies operate the same; in fact, some could qualify for certification without ever having heard of it. Others could qualify for certification, but with numerous conditions to fulfill in order to retain certified status. And still others would require much work in terms of change before they would even be considered for certification.

The reason for these differences, as American psychologist William James noted, is that one's experience is what one agrees to attend to, and only those items one notices shape one's mind. Therefore, those foresters and forestland owners who are most widely aware of the ecological processes that govern their forest are the ones most likely to account for them in forest management and thus are most ready for certification. On the other hand, those foresters and forestland owners who are mostly product oriented and thus see primarily the conversion potential of trees into money are less aware of the ecological processes that govern their forest and hence less ready for certification. We say this up front because what follows is, to the best of our ability, a statement of the problem in its various facets, and is not to be construed as an indictment of the timber industry as a whole. The problem that we will outline is, of course, created through lack of knowledge, cumulative historical events within forestry itself, and, unfortunately, is compounded by those individuals who refuse to change old practices as new knowledge becomes available.

Sustainability

"Sustainable" is a word much bandied about in today's vernacular but often with little clarity of its meaning. Many people seem to think that once something is labeled "sustainable" it is magically fixed, unchanging in time and space. In reality, however, the notion of sustainability, both in concept and practice, is something that cannot be given an explicit definition because its boundaries are elusive, like a horizon of human consciousness that forever recedes as one attempts to discover its limits.[6]

Although our knowledge of the way Nature works has increased dramatically since the dawning of human consciousness, we, as a society, still do not have an answer to one of our most nagging questions: Will the ecosystems of the future, which we are shaping today, continue to function in such a way that the quality of human life will continue, in the best sense, as we have come to expect it? Because we shall never fully know, we continue to pursue the kinds of information that hint at the answer. Meanwhile, the world continues changing in ways we do not expect, cannot predict, and often do not understand. We, in turn, change our ideas and beliefs, albeit slowly and laboriously, as we experience the ever-changing present, which we term chaos when observed events exceed our understanding. In turn, our lack of understanding often becomes an issue of debate.

The term *issue* refers to a point of debate, discussion, or dispute; a matter of public concern; a culminating point leading to a decision. An issue, which becomes a focal point of public concern and debate, such as the biological sustainability of the world's forests, is based on a perceived change in some biophysical circumstance or human value.

This alteration in biophysical circumstance or human value at first necessitates, and finally precipitates, a change in perception, which ultimately precipitates a change in action. It is in dealing with change — in which everything is always becoming something else — that the viability of the future lies today before society, a challenge of personal and social malleability that we must accept, face, and deal with or see society as we know it perish. Herein lies a basic problem of being human; namely, the present is concrete and therefore real, while the future is abstract and thus for most people unreal; or as Steve Egeline of the U.S. Forest Service puts it: "We are managing based on 'far future' predictions, but only perceive and react to 'near future' conditions."

To achieve the balance of biophysical energy and human consciousness necessary to maintain the sustainability of ecosystems, we must focus our questions, both social and scientific, toward understanding the biophysical governance of those systems. Based on such an understanding, as imperfect as it may be, we must, with the long-term welfare of our children in mind, act responsibly in the concreteness of the present so that the abstractness of the future (beyond our lifetime) may prove to be a viable circumstance when today's future becomes our children's present.

Today we have a better chance than at any time in human history of being able to consciously, purposefully affect the future because our technology can give us glimpses farther into the reaches of time than ever before in terms of possible human-caused scenarios based on historic events, current actions, and future probabilities. Despite this, the fact remains that we still do not know enough to make accurate predictions. We think of the future in terms of abstractions and chaos because we lack the ability to make accurate predictions about the outcome of our current actions. Nevertheless, we can increase the probability that we create something meaningful for our children by acting purposefully and consciously in the present with an other-centered (as opposed to self-centered) future outcome in mind — an outcome beyond our lifetime. Then, to maximize the probability of success, we must find the "moral courage" and "political will" to direct our personal and collective human energy toward living within the constraints defined by ecosystem sustainability and our children's long-term welfare and not by short-term political–economic desires.

The systems we are redesigning through our interactions with our environment are continually changing it — all of it, if in no other way than through pollution of the air. Conditions on the North American continent prior to European settlement seem irrelevant because we have little real knowledge of what they were. We can never return to those conditions because there were far fewer people in the world as compared with today. Nevertheless, where knowledge of pre-European settlement does exist, it can act as a template of something that may have worked better than our current ecological systems. The systems we are creating are becoming ever further removed from the types of ecological balances that characterized pre-European conditions. Remember, however, that people were here long before the Europeans, and that here they had already greatly altered the prehuman conditions.[7]

To find an alternative view of today's world, we must find an alternative language — something that works today in terms of the future. To find such a language, says social critic Ivan Illich, we must return to the past to discover the history around which the current "certitudes" were invented, certitudes like "need," "growth," and "development," because these form the organizational core of our modern experience.

Consider, for example, that the language of the "Multiple Use Sustained Yield Act of 1960," although of good intent, is based on an economic assumption totally at odds with ecological reality. As practiced, sustained yield, which is an *economic* concept, means the volume of wood fiber to be cut annually as predetermined by the economic targets of the timber industry and set through Congress. These "hard targets," as they were called until recently, have nothing whatsoever to do with the ecological capability of the forest to produce that volume on a biologically sustainable basis.

Quite the contrary, an economically sustained yield has meant a continuing liquidation cut of available old-growth forest or "excess inventory," as the timber industry has sometimes called it. And when we run out of old-growth forest, as we are, we run out of sustained cut, and thus run out of "sustained yield" — witness the current battles over the last of the old-growth timber in the Pacific Northwest of the U.S.

The very concept of "sustained yield" is founded on the erroneous assumption that society can have a sustained cut of a limitless supply of timber. Here, the philosophical notion is that *a **sustained yield** and **sustainable yield** are one and the same*, which, of course, they are not. To have a biologically sustainability yield that can, in fact, be sustained, foresters must not only understand but also practice forestry is such a way that (1) the rate at which timber is cut is less than or equal to the rate at which it grows, but cannot exceed the growth rate and (2) the concept of biological sustainability must be applied to all acres all the time — not be misconstrued, like industry has done, as a sustained yield by overcutting private lands and then, when the immediate yield is gone, shifting the cut to public lands while the private lands grow more trees for harvest.

Practicing sustained yield by overcutting private lands and then shifting the cut to public lands was, until a few years ago, based on the notion that old-growth forests represented excess standing inventory (industry's terminology), which needed to be liquidated (also industry's term) as fast as possible in favor of productive tree farms. Inherent in this assumption is the erroneous concept that biological processes in the forest remain constant while humans strive to maximize whatever forest products seem desirable.[8] The Central European errors over the past several hundred years, such as converting hardwood forests to coniferous plantations, illustrate well the results of ignoring ecological reality while attempting to maximize short-term profits based on economic assumptions in a system ultimately controlled by biophysical laws.

We cannot, however, manage sustainability for its own sake, because it is thought of in terms of something, such as a sustainable yield of corn, hogs, cattle, or trees. Sustain*able* yield (a *biological* concept) is not the same as sustain*ed* yield (an *economic* concept). Beyond the concept of sustainability, we must recognize that every ecosystem, however defined, is inevitably evolving toward a critical state in which a minor event can, and eventually does, lead to a catastrophic event, which alters the eco-system in some way. The 1988 fires in Yellowstone Park are an example of such a "catastrophe." Given enough time without human interference, however, a system approximates what it was.[7] This ability of the system to retain the integrity of its basic relationships and thereby heal itself is termed "ecological resilience."

Because of the dynamic nature of the evolving ecosystems we attempt to manage, we can only "manage" in terms of an ecosystem's ecological evolution. We cannot manage it just for a sustained yield of products. By this we mean that each acre of forest is constantly changing, so it is not only economically naïve to expect but also physically impossible to force a given acre to produce a predictable volume of timber time after time after time, and so on.

Nevertheless, such predictability is the often-stated objective of many in the timber industry with respect to its economic models as well as its planning models. Because we cannot totally predict the future, however, the only true sustainability for which we can manage is to ensure an ecosystem's ability to adapt to change, such as warming of the global climate and perhaps long-term, human-caused shifts in the biophysical patterns of a landscape.

In other words, if human society is to survive, we must give up our insistence on managing forests by technological harness and economic-political constraint; that is, we must let go of the notion of sustained yield

Photo 1 Wind storms are one of Nature's disturbance regimes. Strong winds often topple weakened trees that, when they fall, leave behind a pit as their roots are pulled from the soil and create a corresponding mound by the uplifted rootwad. When trees blow down over many years, the pits and mounds collectively form what is called a "pit and mound topography" that, along with the fallen trunks, adds a critical structural and functional dimension to the floor of the forest as the wood gradually decomposes and recycles into the soil as part of the nutrient-cycling infrastructure of the forest floor. (Photograph by Chris Maser.)

as overcutting private lands and then shifting the cut to public lands while the private lands are replanted with the next economic crop of trees for harvest. Our only viable choice is to consciously and purposefully work with forests to maintain or repair their resilience in the face of change and the novelty of adaptation — the creative process, which equates to protecting their diversity that in turn equates to sustainability.

Part of this creative process is the spread of Nature's disturbances across a landscape (Photo 1). It is an important process influenced by spatial heterogeneity. Disturbance can be characterized as any relatively discrete event in time and space that disrupts the structure of a population and/or community of plants and animals or disrupts the ecosystem as a whole and thereby changes the availability of resources and/or restructures the physical environment. Regimes of ecological disturbance can be characterized by their distribution in space, size of disturbance, frequency, duration, intensity, severity, synergism, and predictability.[9]

Here it is important to emphasize an often forgotten point. Namely, the greatest, single disturbance to ecosystems has been human disruption of the various regimes of disturbance with which ecosystems have evolved and to which they have become adapted. One of the most obvious examples is the removal of Nature's fire from a forest through fire suppression. Fire is a physical process through which Nature originally designed forests in the Western U.S. But that's not how Gifford Pinchot saw it as he rode through parklike stands of ponderosa pine along the Mogollon Rim of central Arizona in the year 1900.

It was a fine day in June as Pinchot rode his horse to the edge of a bluff overlooking the largest continuous ponderosa pine forest in North America. It was warm, and everything seemed flammable. Even the pine-scented air seemed ready to burn. What a sight! Sitting on a horse in a sun-dappled, perfumed forest without a road to scar the ground, without humanity's machinery to tear the silence, to simply behold such a pristine forest.

"We looked down and across the forest to the plain," he wrote years later. "And as we looked there rose a line of smokes. An Apache was getting ready to hunt deer. And he was setting the woods on fire because a hunter has a better chance under cover of smoke. It was primeval but not according to the rules."[10]

The forest over which Pinchot gazed on that June day in 1900 was three to four hundred or more years old, trees that had germinated and grown throughout their lives in a regime characterized by low-intensity surface fires sweeping repeatedly through their understory. These fires, occurring every few years or so, consumed dead branches, stems, and needles on the ground and simultaneously thinned clumps of seedlings growing in openings left by vanquished trees. Although fire had been a major architect of the parklike forest of stately pines that Pinchot admired, he didn't understand fire's significance in designing the forest.

While Pinchot knew about fire, he was convinced it had no place in a "managed forest." Fire was therefore to be vigorously extinguished, because conventional wisdom dictated that ground fires kept forests "understocked," and more trees could be grown and harvested without fire. In addition, surviving trees, like the ones Pinchot saw in Arizona, were often scarred by the fires, and this kind of injury allowed decay-causing fungi to enter the stem, thus reducing the quantity and quality of harvestable wood.

It was Pinchot's utilitarian conviction about fire's economic evil that became both the mission and the metaphor of the young agency he built, the U.S. Forest Service. Here we must keep Pinchot's two ideas in mind: fire has no place in a managed forest, and what is not used to the material benefit of society is a waste.

In Pinchot's time and place in history, he was correct and on the cutting edge, and the ecological problems caused by such thinking were unbeknownst to him. Nevertheless, incorporation of these ideas into forestry began to take their toll. Only now, decades after the instigation of fire suppression, has the significance of changes in composition, structure, and function of forests become evident.

Recent evidence shows, for example, that some ponderosa pine forests in northern Arizona had only 23 large trees per acre in presettlement times. This presettlement density is in stark contrast to the current density of approximately 850 trees per acre, predominantly small diameters.[11]

Since the advent of fire suppression, there has been a general increase in both the number of trees and the amount of woody fuels per acre. There has also been a decrease in the extent of quaking aspen, which often resprouts from roots following fire, and a corresponding increase in those species of trees that tolerate shaded conditions under closed canopies. And some of these shade-tolerant trees have grown into the forest canopy and formed a ladder of combustible material which a fire can burn from near the ground to the tops of large trees.

It may seem odd, but the ecological degradation of the ponderosa pine forests in northern Arizona in recent times is because of too many trees. Such increased tree density was caused by the introduction of livestock grazing and fire suppression, which shifted the open, parklike, presettlement forests of huge stately trees to dense, closed-canopy stands of less vigorous young trees — an entirely different forest ecologically.

Thus, as we remove Nature's disturbances, we begin to alter a system's ability to resist or to cope with a multitude of invisible stresses to which the system is adapted and adapting through the dynamics of the very regime of disturbance that we removed, such as fire. The precise mechanisms whereby forests cope with stress vary, but one is closely tied to the genetic selectivity of its species. Thus, as a forest changes and is influenced by changing magnitudes of stresses, the replacement of a stress-sensitive species with a functionally similar, but more stress-resistant species maintains the forest's overall functional properties, such as productivity,[12] but at the potential expense of the redundancy of its biological diversity, some of which it has lost.

Forests are also influenced by human-introduced disturbances. In the early 19th century, for instance, there was an emerging view of strong interdependencies among the climate, plants and animals, and the soil, which led to the long-term stability of forests across landscapes. But this notion assumed that the climate remained the same and was considered only in the scale of space. Today, a revised concept is emerging: the spatial patterns, including those of forests, observed on landscapes result from complex interactions among biophysical *and* social forces over time.

As such, the cultural patterns of human use have affected most land-scapes, which become ever-changing mosaics of unmanaged and managed patches of forest and other habitats that vary in size, shape, and arrangement. This spatial patterning is a unique phenomenon on the landscape in any given moment and changes over time.

But forests are ill adapted to cope with some human-introduced disturbances, such as their fragmentation through large-scale clear-cutting. The connectivity of forested areas within a landscape is important to the health of biological processes, as well as to the ability of plants and animals to maintain both viable numbers and viable symbiotic relationships within their habitats. In this sense, the landscape can be considered as a mosaic of interconnected patches of habitat, such as forested riparian areas, which act as corridors or routes of travel between patches of upland forest or other suitable habitat.

It is becoming increasingly apparent, as habitats are fragmented, that the survival of populations of plants and animals in a forested landscape depends on the degree of disturbance, the rate of local extinctions from patches of habitat caused by disturbance, and on how well species can move among patches. Species that are isolated in patches of habitat because of the fragmentation of the landscape have a lower chance of persistence over time than do species that are not so isolated.[13] Fragmentation of habitat is the most serious threat to biological, genetic, and functional diversity and is the primary cause of the present crisis in the extinction of species and the subsequent loss of their contribution to the health and viability of the global ecosystem.

This statement must be taken to heart because strong, self-reinforcing feedback loops characterize many interactions in Nature. Ecosystems comprised of strongly interacting components have long been thought to account for the stability of complex systems. Although these self-reinforcing feedback loops are being increasingly recognized in the intellectual realm of science as important, basic components of ecosystem dynamics, they are too often ignored in daily activities, such as the practice of forestry. At this juncture, however, it is possible to learn a valuable lesson from the global fear that was instilled for more than a year by the Y2K (year 2000) crisis, which never came to pass.

Pause for a moment and consider that no one thought far enough ahead to see what would happen to the intricate feedback loops of which the world's integrated computer system is composed when the millennium changed from 1900 to 2000. The simple changing of two numbers — 19 to 20 — threw much of the world into an expensive, and in some cases potentially dangerous, functional crisis. After all, computers have become our artificial life-support system, and in that capacity are an analog of Nature's life-support system in that both are brittle, overextended, and nonsustainable — the former by an unintentional flaw in design and the latter through continual myopic, self-centered, short-term, linear exploitation. The Y2K crisis momentarily put modern society on notice as never before that we must both face and accept that the Y2K crisis, and by extension those crises in our global environment, which are continually emerging, are issues of appropriate scale with respect to the sustainability of hidden, internal feedback loops. We ignore these pending crises at our growing peril.

Within the constraints imposed by outside forces, a biological system characterized by strong feedback loops is in many respects self-generating. This means that its productivity and its stability are largely determined by its own internal interactions. Put generally, ecosystems that are characterized by strong interactions can be complex, productive, stable, and resilient under conditions to which they are adapted. When critical linkages are disrupted, however, these same systems become fragile and subject to threshold-like changes, which can dramatically affect the biological sustainability of the products we, as a society, value a given ecosystem for and want it to continue producing.

For example, if for some reason the close links between plants and the soil are weakened during a disturbance, alteration of the belowground portion of the ecosystem may lead to poor recovery of the original plant community. This is particularly likely in some coarse-textured soils where mycorrhizal fungi provide much structure by "gluing" soil particles together, thereby providing the capacity of the soil to store nutrients and water.[14] Death of the mycorrhizal fungi due to disturbance from clear-cut logging in such soils leads to further reductions in the populations of belowground mutualists, those organisms that benefit the plants even as the plants benefit them. Feedback loops then rapidly push the ecosystem toward some new state — one seldom to our social liking.

Although stable and resilient against the disturbances that characterize its environment, a system may be exceedingly vulnerable to foreign patterns of disturbance, such as clear-cut logging, fragmentation of habitats, or suppression of fire. The threshold-like disruption of an ecosystem can be avoided to some extent, but only if we understand and protect the critical interactions that bind the diverse components into a whole.

Therefore, modifications of the connectivity among patches of forested habitats can strongly influence the abundance of species and their patterns of movement. The size, shape, and diversity of patches also affect the patterns of species' abundance, and the shape of a patch may determine which species can use it as habitat. The interaction between the processes of dispersal and the pattern of a landscape determine the temporal dynamics of species' populations, but local populations, which can disperse great distances, may not be as strongly affected by the spatial arrangement of patches of habitat as are populations of more sedentary species.

This means that species, both plants and animals, in fragmented landscapes are vulnerable to what ecologist Thomas Ledig calls "secret extinctions" — the loss of locally adapted populations, such as the genetic diversity stored in a local indigenous population of trees that has evolved over centuries or millennia.[15] If locally adapted populations are extirpated, they might never be replaced, and thus their passing inexorably alters the habitat, because other populations of the same species might lack the characteristics necessary to become reestablished in the habitat or might not be able to reach suitable habitat because of major environmental shifts due to the unprecedented speed of the changes brought about by human activities.

Part of the process of maintaining ecological resilience and thus biological sustainability is linked to climate change, the data for which are becoming, in our opinion (mine and Walter's), irrefutable. In the face of our rapidly changing climate, we as a society must increasingly manage landscapes as viable patches of indigenous habitat at the landscape scale. This is a critical concept, because forests migrate as interactive aboveground–belowground communities of symbiotic plants and animals.

To allow forests to migrate in the face of global climate change, as they have for millennia, we must emphasize diversity in all of its aspects, which in turn will do much to ensure the resilience and sustainability of forests, the underpinning of sustainable forestry. We also must pay close attention to the patterns we create on the landscape, because the overall connectivity of those patterns will either allow or inhibit the ability of species and forests to migrate and will determine how much of the genetic variability — stored adaptability — we lose from the gene pool in the "secret extinctions" that accompany the fragmentation of forests and landscapes. Thus, while the current trend toward homogenizing forests within landscapes may make sense with respect to maximizing short-term profits, it bodes ill for the long-term ability of forests and landscapes to adapt to changing environmental conditions on a sustainable basis, which may well endanger the sustainable yield of products we, as a society, require for our survival.

Although we must strive to use forests on an ecologically sustainable basis, such sustainability may prove to be relatively shorter lived than we anticipated in the face of global warming. This being the probable case, we must look beyond our immediate notion of sustaining forests as isolated entities to their long-term sustainability as ever-evolving components of dynamic landscapes, which must be designed to remain adaptable to changing environmental conditions over time.

To this end, an awareness of disastrous consequences, brought about by historically proven unwise choices, could encourage us to change the way we do things so that we may alter a potentially unwanted outcome. But before we can alter the outcome of any historical trend, we must ask fundamentally different questions than have heretofore been asked. After all, an answer is only important when a relevant question has been asked.

Tree farm management, which is based on economically motivated biophysical simplification of complex forests, reigns supreme in the thinking of today's forestry profession in the U.S. despite the rapidly growing body of research that unequivocally demonstrates biodiversity — the diversity of living species and their biological functions and processes — to be an ecological insurance policy for the flexibility of future choice of management options. And because every ecosystem adapts in some way with or without the human hand, our heavy-handedness precludes our ability to guess, much less to know, what kind of adaptations will emerge. We must therefore pay particular attention to ecological redundancy, of which biodiversity is the "nuts and bolts."

Redundancy, as already mentioned, is the ability of more than one species to perform a given function, which strengthens the ability of a system to retain its integrity. Redundancy means that the loss of a species or two is not likely to result in such severe functional disruptions of the ecosystem as to cause its collapse because other species can make up for the functional loss. But there comes a point, a threshold, when the loss of one or two more species will tip the balance and cause the system to begin an irreversible change that may well signal a decline in quality or productivity with respect to our human desires.

Each ecosystem contains built-in redundancies that give it the resilience to either resist change or bounce back after disturbance, but we don't know which species do what or the way in which they do it. When we tinker willy-nilly with an ecosystem's composition and structure to suit our short-term economic desires, we lose both species and their biophysical functions to extinction and thus reduce the ecosystem's biodiversity and hence its redundancies. With decreased biodiversity, we lose choices for desirable actions, which in turn may so alter the ecosystem that it can no longer produce that for which we valued it in the first place. Herein lies the economic fallacies of forestry.

The practice of forestry began with the idea of forests as perpetual producers of commodities. To capitalize on the yield of such commodities as a way of life, the yield had to be economically sustainable, which meant there had to be a disciplined, economic rationale for their exploitation. The "soil-rent theory" became that rationale.

The Soil-Rent Theory

The soil-rent theory — a classic, liberal, economic theory — is a planning tool devised by Johann Christian Hundeshagen in the early 1800s for use in maximizing profits as a general objective of economic activities (an economically sustain*ed* yield as opposed to a biologically sustain*able* yield). It received its final form and mathematical formulation in the mid-1800s through the work of Faustmann and Pressler, and, since its unfortunate adoption by foresters, the soil-rent theory has become the overriding objective for forestry worldwide.[16]

The soil-rent theory is based on the concept of incurring as little financial risk as possible while maximizing profits as much as possible. Financial risk is minimized by selecting the fastest growing species of tree for a given site and then assuming that all ecological variables can be economically converted into constant values, which allows one to calculate, with the illusion of ecological impunity, the age of harvest for the trees (the economically conceived independent variable) that will give the highest rate of return on the economic capital invested.

The criteria used to justify the soil-rent theory are (1) easy establishment of a stand of trees; (2) short rotations, which minimize tying up investment capital; (3) rapid growth, which assures quick profits as much as possible; (4) uniformity of the stands, which assures as much as possible the highest market value for the wood fiber; (5) greatest percentage of usable wood fiber, which again maximizes profits; and (6) continual improvement in the technology for utilization of wood fiber, which also maximizes profits.

The criteria used to justify the soil-rent theory are based on the mistaken notion that ecological variables, such as soil fertility, can in fact be rendered constant through economic rationale and thus ignored in the practice of forestry. Such a notion is erroneous, however, because an economically created independent biophysical variable cannot exist in an interactive living system.

The practice of forestry is therefore based on economic concepts (primarily the short-term maximization of profit) that have nothing whatsoever to do with the ecology or the biological sustainability of forests because there neither is nor can there be constant values and independent variables within an interactive living system. Therefore, today's forestry is merely repeating the mistakes already recorded in the annals of history, the first of which is that "renewable" natural resources are considered to be infinite. The second mistaken notion is that conservation — carrying into the future — of these inherited resources is an economic liability and thus economically unsound. The latter concept is based on the belief that if a 400-year-old tree dies, falls over, rots, and disappears into the soil of the forest floor rather than having been sent to the sawmill and converted into usable boards for building, it is a "waste."

Historical Errors in Forestry

Forestry "management" has been and still is based on at least six flawed assumptions stemming from the soil-rent theory. First is the assumption that *we humans are in control of Nature and thus we can actually "manage" (which means to control) a forest based on the notion of economically perceived constant values and independent variables, which we can either ignore or control over time.* The assumption of management is that simplifying a forest by discarding unwanted parts makes it better and more desirable in terms of society's materialistic demands.[17] We do not "manage" anything, however. We treat the forest in some way and it responds to that treatment as an integrated, interactive living system.

Second is the assumption that once the indigenous forests are liquidated, they can be replaced by tree farms, which not only function better than the original forest but also are forever renewable on a continual plant-cut-plant-cut cycle. *This erroneous notion is based on three further assumptions: (1) an acre of ground that is not growing a desired species of tree at a desired level of stocking is thought to be "unproductive," which is an economic concept — not a biological one* (Photo 2); *(2) that forests*

Photo 2　Red alder, which is still considered by many foresters to be a nonproductive "weed" to be eliminated when possible in order to plant monocultures of the economically important Douglas fir. (Photograph by Chris Maser.)

and tree farms are one and the same, which they are not; and (3) that the mere act of planting trees in neat, economically motivated crop-like rows is the same as Nature's creative process of growing a diversified forest — *hence the term "reforestation."* Here the problem is that our thinking and therefore our models are linear, while the cyclical forest occupies a sphere vastly different from both our thinking and our models.

Third, we either fail to realize or refuse to accept that all our "management" is directed toward what we see aboveground and that we cannot alter the aboveground without simultaneously altering the belowground. *Here the mistaken thinking is that what is done aboveground in the name of forest management, other than compaction of the soil, has little or no effect on how belowground biophysical processes function in a forest.* We do not, therefore, even think about or plan for belowground processes. Thus we either fail to understand or refuse to accept that each tree, each stand of trees, or each forest is a mirror reflection of the soil's ability to grow that particular tree, stand, or forest just once!

Fourth, forest productivity rests on five biophysical factors:

■ The depth and fertility of the soil in which the forest grows
■ The quality, quantity, and timing of the precipitation reaching the forest
■ The quality of the air infusing the forest
■ The amount and quality of the solar energy interacting with the forest
■ The stability of the climate in which the forest grows

By failing to include these five factors in our economic and planning models and thus make the inaccurate assumption, through omission, that these factors are constant values. Even if we try to build them into our ecological models, we do so in a linear mode, which is the only way we know how. We therefore think and act as though soil, water, air, sunlight, and climate are constant in their behavior and that the only variable we manipulate is the kind of tree we plant and its rate of growth.

These biophysical factors are *all variables*, however. Soil, for example, is eroded in two ways, chemically and physically. We do both. We pollute the water and the air with chemicals. Air pollution directly affects the forest by altering the quality of the soil and the water as well as the quality and quantity of the sunlight that drives the forest processes. The chemicals we dump into the air also alter the climate and thus the environment in which the forest grows. The quality and quantity of soil, water, air, and sunlight with which a forest interacts and on which a forest is interdependent are all variables that in turn are dependent on the stability of the climate, itself a variable, and must be so treated.

The fifth error is the notion that biodiversity is counterproductive to long-term economic gain and is thus dispensable. *The false idea in such thinking is that diversity — biological, genetic, and functional — can be disregarded and thus forests can be simplified with ecological impunity.*

With this mindset, foresters purposefully simplify the only part of the forest we see, that which is aboveground — an error made as a short-term economic expedient toward the primary objective of maximizing the production of wood fiber. In this sense, the forest industry not only eliminates as much biological diversity as possible, such as economically undesirable species of plants and animals, but also eliminates as much "economically undesirable" genetic diversity as possible.

This is deemed good forestry, because mainstream thinking either does not realize or refuses to accept that the forest one sees aboveground is a faithful reflection of the health of the forest hidden belowground in the soil — the biological processes and their ability to grow a particular stand of trees. *Here the erroneous concept is that nothing in a forest has value unless it is converted into something else, i.e., trees into boards, boards into houses, and so on — economic conversion potential. Put a little differently, nothing in a forest has intrinsic value, only extrinsic value.*

The sixth error is clinging to the idea that an economic endeavor must be ever-expanding to be healthy. Thus we attack the world's renewable natural resources from an ecologically exploitive point of view as though there were no tomorrow, which produces increasingly finite limits on most, if not all, "renewable" resources.

The timber industry, particularly in areas where considerable indigenous forest remains, operates in a perpetual expansionistic mode, and as a result, the world's forested resources are rapidly, often irreplaceably shrinking. *The inaccurate concept in this kind of thinking is that indigenous old-growth forests (which have in the past been called "biological deserts," "cellulose cemeteries," and "bug-infested jungles" by industry) must be logged in order to be valued or they will simply go to waste because they are "over mature" — a strictly economic concept to discount the intrinsic value of old-growth forests.* Perpetual expansion, however, involves liquidating the indigenous forest and ultimately exhausting the soil.

The forest's death knell is being sounded by the ever-increasing push for more and more intensive tree farm management on more and more acres based on linear, industrio-economic thinking. So the question becomes one of whether an economically sustained yield is — or ever was — possible in the face of advancing industrialization and the exploding human population, both of which threaten to so pollute and exhaust the soil that the collapse of ecosystems is increasingly probable. *This is a notion vigorously denied by those in the forestry profession who are not only encrusted in the old simplified, stereotyped reductionist mechanical world view, but also practice informed denial as the best way of dealing with unwanted information.*

The Reductionist Mechanical World View

"We can no longer assume that nature's services will always be there free for the taking," writes Janet N. Abramovitz, senior researcher at the Worldwatch Institute.[18] Despite our best efforts to think ahead, we will rarely, if ever, be able to ascertain the full impact of our actions, the consequences of which for Nature will most often be unforeseen and unpredictable. The loss of species and habitats and the degradation and simplification of ecosystems can and usually do impair Nature's ability to provide the necessary services we depend on for life. "Many of these losses are irreversible," says Abramovitz, "and much of what is lost is simply irreplaceable."

This kind of thing, losses from and disruptions to the ecosystem, is not supposed to happen, according to the reductionist mechanical world view, which is today overlain by the notion of continual economic expansion. The reductionist mechanical world view is based on the notion that the economic process of producing and consuming material goods and services has no deleterious effects on the ecosystem because the assumption is not only that natural resources are limitless but also that any unintended effects of the economic process, such as pollution and environmental degradation, are inconsequential.[19]

In contrast to the dominant world view, the paradigm of sustainability is neither mechanical nor reversible; it is entropic, which means that the Earth's resources and its ability to absorb and cleanse the waste produced by humanity's economic activities are both finite, as Distinguished Professor Nicholas Georgescu-Roegen of Vanderbilt University points out.

Georgescu-Roegen goes on to explain that it is easy to confirm the Entropy Law by simply observing that a lump of coal or a piece of wood, which has been converted into heat energy through burning, cannot be reconstructed from the resulting carbon dioxide and solid residues. The Entropy Law, or Second Law of Thermodynamics, states that the process of converting mass, such as a lump of coal or a piece of wood, into heat energy is irreversible. Although Georgescu-Roegen knew that the increase in entropy and the corresponding decrease in available energy are absolute over time, he thought that people could live sustainably over the long run if they used natural resources in the most renewable manner possible. By tying the economic process to the entropy of the physical world, Georgescu-Roegen, the economist, was pointing out that for economic thought to be viable over time required the common sense of a shift from the old reductionist mechanical world view to a paradigm built around sustainability.

Only when one understands how deeply rooted in our Western industrialized society the reductionist mechanical world view (with its overlay of economic expansionism) really is will one begin to appreciate just how difficult it will be to change our Western culture from a destructively exploitive way of life to a more sustainable approach to life. The following discussion will give you some appreciation of how long ago our imperial concepts of economics and Nature began to evolve.

As you read the following critique of the rationalistic thinkers, bear in mind that they gave us not only such things as science, calculus, etc., but also the Renaissance, which helped to guide us out of the Dark Ages. In addition, much of what they gave us is still valid, and that which is not forms, in large part, the foundation of our current state of knowledge.

Rationalistic thinkers, such as Francis Bacon (1561–1626, philosopher, essayist, and former Lord Chancellor of England), Galileo Galilei (1564–1642, Italian scientist and philosopher), René Descartes (1596–1650, French philosopher and mathematician), John Locke (1632–1704, English philosopher), Isaac Newton (1642–1727, English mathematician, scientist, and philosopher), Carl von Linne (1701–1778, Swedish botanist), and Adam Smith (1723–1790, Scottish political economist and philosopher), legitimized and institutionalized the lust for and imperial acquisition of material wealth over which feudal society had fought for so long. In so doing was born the reductionistic mechanical world view.

Consider the collective paradigm of these renowned men: Nature's sole value is in service to the material desires of humanity; in turn, humanity's role is to dominate Nature to the greatest extent possible through the aid of science-based technology (Bacon, Linne, Locke). Nature, which is simply a storehouse of raw materials for human use, must be tortured before its secrets will be revealed, but once revealed, the world can be rendered into a paradise through science and management by humans because the world is made for humanity, not humanity for the world (Bacon, Smith). Once wrested from Nature, only those secrets that are measurable and quantifiable are real or relevant and can be studied (Galileo). Because real things are both measurable and quantifiable, they must operate through predictable linear mechanical principles, like an enormous machine (Descartes, Newton). And like a machine, real things can be understood by disassembling the things themselves into smaller and smaller, more manageable pieces, which can then be rearranged in an order deemed logical to the human mind (Descartes).

With reductionistic mechanical logic, major segments of Western industrialized society confer upon themselves the unlimited rights of individual private property (Locke) for which people must compete with one another in pursuit of their own self-interests (Smith). Such self-interest is to be free from any government interference because the "invisible hand" of moral guidance will temper self-interest in the pursuit of material wealth — for the betterment of society (Smith). While Smith's "invisible hand" may have spiritual connotations, they are ignored in the current pursuit of self-interests in the form of material wealth. Further, his notion of a Higher Moral Principle guiding human action was already overshadowed by the accepted reductionistic mechanical posits of Bacon, Galileo, Descartes, Locke, and Newton.[20]

Now add to this paradigm the soil-rent theory, which puts in place another piece of reductionistic thinking as promulgated by much of today's timber industry as it practices forestry. Add to the soil-rent theory the notion of intensive utilization of wood fiber, and another piece of reductionistic thinking falls into place, as exemplified by the thoughts of Clyde Martin of the Western Pine Association, who wrote in the *Journal of Forestry* in 1940 that "Without more complete and profitable utilization we cannot have intensive forest management.... When thinnings can be sold at a profit and every limb and twig of the tree has value, forest management will come as a matter of course."[21]

In response to increased competition in today's market place and to the growing number of environmental restrictions, private forest landowners, particularly timber companies, are seeking to increase the yield of wood fiber on tree farms with very short rotations. Because the intensive silvicultural practices employed to increase the yield resemble those used in intensive agriculture, it is believed that herbicides are necessary to boost the volume of wood fiber. These intense silvicultural practices, which Clyde Martin could only dream about in the 1940s, can boost the volume by 128 percent and the rate of economic return by 12 percent. "The high yields possible from fiber farming could allow changes in land use, from timber production to other uses, while maintaining supplies of low-cost fiber."[22] Perhaps it could change the potential use of land, but which timber company is going to forgo a potential increase in profit of 12 percent to a competitor? "The real question is whether we can adopt new technologies to produce more wood from plantation [tree farm] forestry."[22]

Then add clear-cutting, which allows the conversion of biologically complex forests to biologically simplistic, intensively manicured monocultures in the form of economically designed tree farms — which are *not* forests. In terms of tree farms, this thought process follows the current economic paradigm of intensive agriculture, which is based on reductionistic thinking and thus has much to do with short-term economics and little to do with the long-term biological sustainability of such forest components as soil. And of all the forest components, soil is the most alive, the most complex, and the most ignored in forestry.

Next, consider the battery of herbicides used to kill or reduce any unwanted vegetation that is perceived to compete with the economic timetable for the potential harvest of crop trees. The ecological function of the unwanted vegetation, which benefits a given site in many hidden and unknown biological ways, is seldom taken into account. After all, a tree farm is no longer a forest and thus fits into the economics of the soil-rent theory, which further compounds the reductionistic approach of modern forestry.

More recently, cloning and genetic engineering of economically desirable species of trees to produce economically desirable behavioral traits, such as fast growth, for strictly economic yields add yet another piece to forestry's reductionistic tool kit. The rationale behind these genetically engineered trees is as simplistic as selectively breeding beef cattle to gain more weight faster to bring a higher price in the market place as soon as absolutely possible. Thus, the tree farm becomes ever more important economically.

The reductionistic mechanical world view has led to the mind/body split and the human/Nature separation and is ever increasing our sense of isolation from one another and from Nature. This dualism has led us to treat Nature as a commodity from which we are independent and separate. By separating ourselves from Nature, we have justified our trying to control the uncontrollable.

Our analytical perspective involves a four-part process: (1) disarticulate the system into its component parts, (2) study each part in isolation, (3) glean a knowledge of the whole by studying its parts, and (4) rearrange the parts in a way that fits the logical to our reductionistic mechanical world view. This is like disarticulating a live cat, rearranging its parts, putting them back together again, and expecting the cat to live and function as before.

The implicit assumption of our analytical perspective is that systems are aggregates of interchangeable parts that function in a linear fashion. Thus by optimizing each part, we optimize the whole. We continually fragment our problems into smaller, more "manageable" (albeit linear and increasingly dysfunctional) pieces while our social and environmental challenges are increasingly interlinked and systemic.

Today, fragmentation, which looks at the parts and ignores the whole, continues to disintegrate our social structure by obliterating the sense of a society as a living system. Fragmentation as practiced today, such as professional specialization, special interest groups, and political lobbyists, is the very foundation of professionalism, and yet it is making our society increasingly ungovernable. The triumph of such reductionistic thinking has given rise to a whole set of conditions under which we try to operate in isolation from the system itself.

This kind of fragmentation led quantum physicist David Bohm to say, "Starting with the agricultural revolution, and continuing through the industrial revolution, increasing fragmentation in the social order has produced a progressive fragmentation in our thought."

Our social predicament, including forestry, is a legacy of our reductionistic mechanical world view, which finds value only in those material things that can be measured and quantified and discounts all things defying material valuation. A result of such narrow, rigid linearity is one of the most insidious patterns of Western industrialized economic thought: that whatever the immediate focus of one's attention might be — trees, cattle, people, corn, or soil — it has no value unless and until it is converted into money. Nothing, according to our economic system, has intrinsic value. Not even money itself.

The only value of anything seems to be its "conversion potential." Conversion potential is oriented almost completely toward the control of Nature and the conversion of natural resources into economic commodities as fast as possible. Conversion potential dignifies with a name the erroneous notion that Nature has no intrinsic value and must be converted into money before any value can be assigned to it. All of Nature is thus seen only in terms of its conversion potential.

Although Western civilization has long followed the thinking of Isaac Newton that the universe operates in a predictable manner, like an enormous machine, we now know his premise to be incorrect. A new vision of a single organic whole (such as a forest as opposed to individual trees) is being derived through the revolution in physics, primarily quantum mechanics and the work of Albert Einstein et al. But the thinking of many people in the forest industry has yet to catch up with the knowledge of modern physics and a changing world view.

Having been long steeped in the reductionistic mechanical world view, forestry too easily dismisses as impractical idealism any attempt to refocus from bread-and-butter issues to ideas and processes. But it is only a matter of time before a shifting social consciousness will force the timber industry to accept a change in its thinking. This will be a change from its current linear thought process to a holistic, systems approach, where the indicators of health are rooted in the quality of the relationships between and among the parts within a single system and among systems.

This shift in thinking means recognizing the value of relationships and accepting that the only way anything can exist is encompassed in its interdependent and interactive relationship to everything else, which means there is no such thing as an ecological constant or an independent variable. As such, every relationship is dynamic, constantly adjusting itself to fit precisely into all other relationships, which consequently are adjusting themselves to fit precisely into every other relationship, and so on *ad infinitum.*

Can so fluid a notion as ever-adjusting relationships be made to work within our current, rigid, reductionist, mechanically oriented social construct? No, because through self-reinforcing behavioral feedback loops, our present social paradigm condemns change as a condition to be avoided at almost any cost. Nevertheless, the perceived security we have so long sought through ever-increasing consumption, such as forestry, militaristic technology, and domination over Nature, has actually threatened our long-term, social survival.

This threat is quickly approaching, if it is not already here, which prompted Czech President Vaclav Havel to observe: "Without a global revolution in human consciousness a more humane society will not be possible." Everything changes with time, however, and today we have a crisis in perception because our behavior and the extension of our feelings, thoughts, and values are increasingly apparent as the root of our manyfold social/environmental problems. These problems are the outworking of a single, overarching crisis of spiritual/moral values brought about largely by the clinging of industrialized societies to a self-centered, mechanical reductionist world view and its associated value system of economic expansionism.

Under the influence of a reductionistic mechanical world view, which is today heavily overlain with a demand that our economic system be ever-expanding, it is too easy to dismiss as impractical idealism any attempt to refocus from immediate political issues to long-term processes and futuristic ideas. Further compounding the belief that long-term processes and futuristic ideas are merely impractical idealism is the notion of conversion potential. For many people, as we have already pointed out, the only value of anything is its "conversion potential."

No matter what central issue is discussed, therefore, the dynamics are the same — an underlying crisis of perception. Our continued acceptance of a reductionist mechanical world view glued together by the notion of conversion potential as the absolute truth and the only valid way to knowledge has led to the current global crisis of deforestation worldwide.

Now, as history's veil enshrouds the events of the 20th century, the cherished cultural values of the reductionistic mechanical world view are in deadly grapple with the revelations of science that increasingly challenge that view. One of the major problems facing us today is the way in which we accept that challenge, be it from a product frame of reference (however the "product" is defined) or that of a systems approach, and all the shades in-between. The differing perspectives define the terms of debate taking place within the linearity of thinking within the profession of modern forestry.

The Linearity of Current Thinking

By assuming that all ecological variables can be converted into economically constant values (which thus constitute ecologically constant values), one is misled to think that all one must do to have an economically sustainable tree farm is calculate the species of tree, the rate of growth, and the age of harvest that will give the highest rate of economic return in the shortest time for the amount of economic capital invested in a given site. Such a notion is based on the linearity of thinking in the forestry profession.

The concept of linearity reflects the doctrine of progress, which is represented by the European invaders of the New World, who crossed oceans and continents annexing everything in their path for material gain. They used the land and abused it and moved on, for there always seemed to be another hill to go over, a virgin resource to discover.[7] Now, with nowhere else on Earth to go, we are moving into space on a course of no return, a course that is presumed to bring us into a human-made, material paradise unfettered by Nature's laws.

Within this concept, modern society moves through time in the same way as did the Europeans who invaded the New World, discarding old experience as the new is encountered. Thus, we seldom learn from history because in our minds we never "repeat" the old mistakes, but in reality, we repeat them constantly. We deny our repeated mistakes, however, in our blind drive for material progress because there is but one definition of progress in this vision — an endless galactic journey of discovery, conquest, and exploitation. To return is to come back to the used and the discarded, when progress is to forever exploit the new.

Linearity contains the notion that anything is justifiable so long as and insofar as it is immediately and obviously good for something else — its economic conversion potential. What, we ask, is it good for? What can it be converted into? And only if it proves to be immediately good for something other than its intrinsic value are we ready to raise the question of its "real" value: How much money is it worth? We ask because since it can only be good for something else, obviously it can only be worth something else.

Because linearity covets only conversion potential, it discounts intrinsic value in everything it touches. Excellent examples of such linear thinking in the environmental sense are the endangered species. Because linear vision looks fixedly straight ahead with the notion that for an economic endeavor to be healthy it must be ever-expanding, it necessarily views any species that hinders such expansion as expendable. Further, if a species cannot be converted into something else, it has no value: consider the Pacific yew tree — thought to be a "weed" before it became the only source of taxol, a drug used to treat breast and ovarian cancer in women.

Linearity never looks back. Its premise: there can be no return. Linearity is above all the doctrine of possession, which is uncomplemented by the doctrine of relinquishment or sharing.

Processes are invariably cyclic, rising and falling, giving and taking, living and dying in space on ever-expanding ripples of time. Yet linearity places its emphasis only on the rising phase of the cycle — on intellectual knowledge, production, expansion, possession, youth, and life. It relentlessly shuns intuition, return, idleness, contraction, giving, old age, and death.

Waste is thus a concept that can only be born from a vision of economic linearity. According to this notion, every human activity produces waste, because every human activity is linear.

The cyclic vision, on the other hand, ultimately sees our life as an endless repetition of basic and necessary patterns in the circular dance of use and renewal, expansion and contraction, life and death. This vision sees everything as interdependent, as fulfilling the ecological excellence and uniqueness of its function, which means there is no such thing in Nature as an "independent variable." Everything in the Universe is patterned by its interdependence on everything else, and it is the dynamic pattern of interdependence that produces the novelty of change — the universal constant with which we must interact and through which we grow.

A forest is a cycle of interdependent processes in relation to time, completing its cycle only in the memory of many human generations. We seem to ignore this, however, because all of our models — economic, managerial, and even ecological — are short-term and linear to fit within the memory of one's own lifetime and the linear construct of one's thinking. We chose them to be so because we don't have the capability to construct them in any other way.

Linear models can only predict in a straight line in the very short term, but the cyclical nature of the forest touches that line for only the briefest moment in the millennial life of the soil, the womb from which the forest grows. And yet, in that instant, with grossly incomplete, short-sighted knowledge and unquestioning faith in that knowledge we cast the sustained-yield prediction of all of our tree farm management into the unforeseeable future.

We then liquidate the old-growth forest, convert it into a grossly simplified economic tree farm that we think will be forever rapidly growing and that we predict will have a magical sustained yield, even as we ignore biological, genetic, and functional diversity. When someone challenges the conversion of forests into simplistic monocultural tree farms, the timber industry inevitably asks: Do we really need such a variety of species? If so, what do they contribute to a local forest or to the world's ecosystem?

A variety of species increases the stability of ecosystems by means of feedback loops, which as previously stated are the means by which biological processes reinforce themselves. The complementary biophysical function performed by each species within an ecosystem (the feedback loop) is what makes each individual species so valuable. Thus, each species by its very existence has a shape and therefore a structure that in turn allows certain functions to take place, functions that interact in a complementary fashion with those of other species. All of this is governed ultimately by the genetic code, which by replicating species' character traits builds a certain amount of redundancy into each ecosystem.

Although an ecosystem may be stable and able to respond "positively" to the disturbances in its own environment to which it is adapted, this same system may be exceedingly vulnerable to the introduction of foreign disturbances to which it is not adapted. We can avoid disrupting an ecosystem's feedback loops only if we understand and protect the critical interactions that bind the parts of the ecosystem into one.

Diversity of plants and animals therefore plays a seminal role in buffering an ecosystem against disturbances from which it cannot recover. When we lose species, we lose their diversity of structure and function and their genetic diversity, which sooner or later results in complex ecosystems becoming so simplified they will be unable to sustain either themselves or us. Any societal strategy aimed at protecting diversity and its evolution is a critically essential step toward ensuring an ecosystem's ability to adapt to change.

If we fail to adequately protect biological, genetic, and functional diversity in our drive for economic gain, whatever else we may accomplish will be moot. This statement applies to all ecosystems, not just forests. The bottom line is that diversity counts, and we must of necessity protect it at any cost, which brings us to biological capital vs. economic capital.

Biological Capital vs. Economic Capital

An indigenous old-growth forest has three prominent characteristics: large live trees, large standing dead trees or snags, and large fallen trees (Photo 3). The large snags and the large fallen trees, which are only altered states of the live old-growth tree, become part of the forest floor and are eventually incorporated into the forest soil, where myriad organisms and processes make the nutrients stored in the decomposing wood available

Photo 3 The three components of large woody debris are seen in this photograph. To the right is a large, live, old-growth Douglas fir. In the bottom foreground is a large fallen Douglas fir reinvesting its biological capital into the soil, and between and behind the two dark trees to the left of the photograph is a large standing dead tree or snag. While the snag is standing, it houses cavity-nesting birds, as well as bats and flying squirrels that use the abandoned cavities. When the snag falls, it, like the fallen tree, reinvests its biological capital into the soil and the next forest. (USDA Forest Service photograph.)

to the living trees. Further, the changing habitats of the decomposing wood encourage nitrogen fixation to take place by free-living bacteria. (Nitrogen fixation is the conversion of atmospheric nitrogen to a form usable by living organisms, such as a tree.) These processes are all part of Nature's rollover accounting system, which includes such assets as large dead trees, biological diversity, genetic diversity, and functional diversity, all of which count as reinvestments of biological capital in the growing forest.

Intensive, short-term, tree farm management disallows reinvestment of biological capital in the soil and therefore in the forests of the future because such reinvestment has come to be erroneously seen as economic waste. We in Western industrialized society therefore plan the total exploitation of any part of the ecosystem for which we see a human use, and we plan the elimination of any part of the ecosystem for which we cannot see such a use. With this myopic view, we have created the intellectual extinction of Nature's diversity through our social planning system, which inevitably leads to biological extinction of species and their functions within the forest ecosystem.

After the indigenous forest is liquidated, we may be deceived by the apparently successful growth of a first tree farm, which lives off the stored, available nutrients and processes embodied in the soil of the liquidated indigenous forest. Without balancing biological withdrawals, investments, and reinvestments biological interest and principal are both spent and so both biological and economic productivity must eventually decline.[8] The dysfunctional "managed forests" (= tree farms) of Europe — biological deserts compared to their original forests — bear testimony to such shortsighted, economic folly.

Converting a mature forest into a monocultural tree farm and/or applying fertilizer to an existing monocultural tree farm are neither bio-logical *re*investments nor economic *re*investments in either the forest or the soil; they are economic investments in "crop trees!" The initial outlay of economic capital required to liquidate the inherited forest, plant seed-lings on bared land, and fertilize the young stand is an economic invest-ment in the intended product. But a forest does not function on economic capital. It functions on biological capital — the decomposing wood of large fallen trees (Photo 4 and 5) (as well as biological diversity, genetic diversity, and functional diversity[6,23]) — which brings into question the economic practice of salvage logging.

Salvage Logging

In the mainstream of today's consciousness in the forestry profession, salvage logging to capture perceived economic waste in the forest is seen as a viable, even necessary practice to maintain the flow of wood fiber from dwindling forests, especially old-growth forests (Photo 6). There are, however, two false assumptions in the industrial concept of waste: (1) that to be of value, any potential commercial forest product (such as a tree with usable wood) must in fact be used by humans or it is wasted (Photo 7); and (2) because any commercial product not used by humans is a waste, renewable natural resources carried into the future are economically discounted, which is de facto discounting the generations of the future in favor of immediate economic gain.

Be that as it may, there is no such thing as biological waste in a forest. But the economic concept of waste, which discounts nonmonetary social values and all intrinsic ecological values, has spawned the industrial concept of salvage logging, usually in the form of clear-cutting. *Here, the erroneous thinking is that clear-cut logging mimics fire or any other of*

Photo 4   Fallen trees are a critical part of the forest in that they not only create habitat, hold soil in place, act as reservoirs of water in summer, and are sites of nitrogen fixation and storage but also are a ferment of biological activity as they decompose and reinvest their borrowed elements in the soil to rise again in trees of the future. (USDA Forest Service photograph by Kermit Cromack.)

Nature's disturbance regimes and is therefore ecologically defensible. Clear-cutting, however, is only an economic expediency in which we find no biological justification because it mimics nothing in Nature and is therefore ecologically untenable (Photo 8).

Instead of focusing on the erroneous notion of waste to justify salvage logging for any reason, foresters must learn that what they conceive as economic waste is, in reality, biological capital (large snags, large fallen trees and logs, including some large merchantable trees) that is essential for forest health. In turn, biological capital must be reinvested in the forest if its intrinsic value is to be realized. Biological capital, in the form of large organic material, helps to maintain soil health, which in large measure equates to forest health. Forest health, in turn, often equates to the long-term economic health of the timber industry. ("Reinvest" means to leave, to forego some potential short-term profits in the forest in the form of merchantable trees — both live and dead — to assure, for future generations, long-term soil fertility and thus the long-term productivity of the forest.)

Photo 5 **A large fallen tree that is fragmenting with seasonal wetting and drying, freezing and thawing. As it breaks down, it forms a "nurse log" to western hemlock seedlings that find it an ideal growing medium. Its spongy wood, through which the seedlings' roots can easily grow, acts as a reservoir of water and has a high content of readily available nutrients. (Photograph by Chris Maser.)**

Planting and fertilizing trees are investments in the next commercial stand, not reinvestments. They are investments in a potential product. We in Western industrialized society do not reinvest in maintaining the health of biological processes, such as those in the soil, because we see only an isolated commercial product (the trees), not the forest as an integrated living system. We do not reinvest because we see only the conversion potential of the trees into boards for building or pulp for paper, not their intrinsic value in the forest. We do not reinvest because we insist that ecological variables are really the constant values of the economic realm, which we need not consider.

What, you might ask, will really be lost if forests are converted into economic tree farms through such practices as clear-cutting and salvage logging? Before we (Walter and I) can answer this question, you must understand that all things in Nature's forest are neutral when it comes to any kind of human valuation. Nature has only intrinsic value. Thus, each component of the forest, whether a microscopic bacterium or a towering 800-year-old tree, is allowed to develop its prescribed structure, carry out

Photo 6 Under the notion of "salvage," the large fallen tree in the center of the photograph, which is lying under the fallen snag, would be considered a waste if it were not cut into logs and taken to the mill to be converted into lumber. The concept of "salvage," which precludes the idea of intrinsic biological value, is a strictly utilitarian concept based on immediate economic return and says, in effect, that any usable tree left in the forest is an economic waste. (Photograph by Chris Maser.)

its prescribed function, and interact with other components of the forest through their prescribed interdependent processes and feedback loops. No component is more or less valuable than another; each may differ from the other in form, but all are complementary in function.

Consider, for example, the coniferous forests of the Pacific Northwest in which Douglas fir and western hemlock predominate in the old-growth canopy. Herein lives the spotted owl, which preys on the northern flying squirrel as its stable diet. The flying squirrel, in turn, depends on truffles, the belowground fruiting bodies of mycorrhizal fungi. (The term mycorrhiza, meaning "fungus-root," denotes the obligatory symbiotic relationship between certain fungi and plant roots.) Flying squirrels, having eaten truffles, defecate live fungal spores onto the forest floor, which, upon being washed into the soil by rain, inoculate the roots of the forest trees. These fungi depend for survival on the live trees, whose roots they inoculate, to feed them sugars, which the trees produce in their green

Photo 7   A merchantable fallen old-growth Douglas fir blown over by the wind, which, left alone, will decompose and recycle into the soil as a reinvestment of biological capital from the present forest through the soil into the forests of the future. The man standing next to the tree is 6 feet, 7 inches tall. (Photograph by Chris Maser.)

crowns. In turn, the fungi form extensions of a tree's root system by collecting minerals, other nutrients, and water that are vital to the tree's survival. Mycorrhizal fungi also depend on large rotting trees lying on and buried in the forest floor for water and the formation of humus in the soil. Further, nitrogen-fixing bacteria occur inside the mycorrhiza, where they convert atmospheric nitrogen into a form that is usable by both the fungus and the tree. Such mycorrhizal/small mammal/tree relationships have been documented throughout the coniferous forests of the U.S. (including Alaska) and Canada. They are also known from forests in Argentina, Europe, and Australia.

All this is complicated, but so is an indigenous forest. To add to the overall complexity, a live old-growth tree eventually becomes injured and/or sickened with disease and begins to die. How a tree dies determines how it decomposes and reinvests its biological capital (organic material, chemical elements, and functional processes) back into the soil and eventually into another forest.

A tree may die standing as a snag to crumble and fall piecemeal to the forest floor over decades, or it may fall directly to the forest floor as

Photo 8 Clear-cutting, like that done on a commercial basis, is strictly an economic expediency and mimics nothing in Nature. In short, clear-cutting is *not* sustainable forestry. (USDA Forest Service photograph.)

a whole tree. Regardless of how it dies, the snag and fallen trees are only altered states of the live tree; the live old-growth tree must therefore exist before there can be a large snag or fallen tree.

How a tree dies is important to the health of the forest because its manner of death determines the structural dynamics of its body as habitat. Structural dynamics, in turn, determine the biological-chemical diversity hidden within the tree's decomposing body as ecological processes incorporate the old tree into the soil from which the next forest must grow. What goes on inside the decomposing body of a dying or dead tree is the hidden biological and functional diversity that is totally ignored by economic valuation. That trees become injured and diseased and die is therefore critical to the long-term structural and functional health of a forest.

The forest is thus an interactive, organic whole defined not by the pieces of its body, but rather by the interdependent functional relationships of those pieces in creating the whole — the intrinsic value of each piece and its complementary function. These functional relationships are totally ignored in salvage logging.

Let's return for a moment to the Pacific Northwest and consider that the spotted owl preys on the flying squirrel, which depends on truffles for its diet. The fungus, of which the truffle is the reproductive part, is closely associated with large wood on and in the forest floor. The squirrel, the owl, and the fungus all depend on the same wood!

Salvage logging disrupts this interdependent relationship by removing all merchantable dying and dead trees. The danger of such logging lies primarily in its philosophical underpinnings that justify immediate economic considerations to the exclusion of all else, including ecological feedback loops that are vital to the health of the forest. Salvage logging, as currently practiced, has these immediate consequences:

- Areas where logging has heretofore been prohibited, i.e., roadless areas, will be opened to roads thus destroying forever their integrity as roadless areas.

- Arson fires will probably increase to stimulate salvage sales as a means of logging as much of the remaining old-growth forest as possible.

- Timber will most likely be salvaged through clear-cutting, which is a drastic biological simplification of a complex forest ecosystem.

- Salvage, as normally practiced, mimics clear-cutting, and clear-cutting mimics nothing in Nature.

- Normal logging is designed to make money, but within at least some planned ecological constraints. Salvage logging is reactive to keep from losing possible monetary gains and is thus unplanned, opportunistic, and without ecological constraints.

- Normal logging compacts soil and removes a preestablished volume of timber, theoretically within some ecological constraints. Salvage logging is a reentry of logged sites, which further compacts the soil, nullifying any ecological constraints.

- Initial logging is most often based on what is and is not to be cut with at least some consideration of ecological contraints. Salvage logging, on the other hand, opens the real possibility of individual, economically driven carte blanche on-site interpretation of what to cut without regard to ecological considerations, including such things as live "risk trees" or live "associated trees."

We do not question cutting some dying and dead trees because there is nothing morally wrong or ecologically harmful in so doing. But we do question the practice of salvage logging per se. The outcome of such ecological folly, in the name of short-term economic profitability, is inevitably passed on to the children for generations to come.

Salvage logging, one of the most ecologically dangerous practices in modern forestry, employs overriding short-term economic rationale as an excuse to summarily ignore all current ecological knowledge about the long-term biological sustainability of forests. The sole objective of salvage logging, particularly in the Pacific Northwest, is to convert trees into money, thus replacing the art of forestry with the technology and economics of cutting trees.

Salvage logging epitomizes traditional forestry, which is the myopic economic exploitation of trees at the supreme cost of the biological health of the forest as a living system. In turn, traditional forestry focuses on growing and then cutting trees as rapidly as possible to maximize short-term profits.

The potential for converting trees and other resources into money counts so heavily because the economically effective horizon in most economic planning is only about five years away. Thus, in traditional linear economic thinking, any merchantable tree that falls to the ground and reinvests its nutrient capital into the soil is considered an economic waste, i.e., it has not been converted into boards for human use and hence into money.

In forest management, the notion still predominates that anything without monetary value has no value and anything with immediate monetary value is wasted if left unused by humans. Short-term economic profitability is thus always the goal and is politically justified by such maneuvers as attaching salvage logging as a rider to an unrelated Congressional bill that the President could not politically afford to veto. Unfortunately, the long-term ecological–economic price of such maneuvers will be paid for by the generations of the future.

Clearly, we must change our thinking and our actions with respect to how we treat forests in the present for the long term. Forests, after all, are not solely the endless producers of economic commodities and amenities they heretofore have been assumed to be. They are instead interactive living systems controlled by nonnegotiable ecological constraints that, when ignored, can destroy the system as we know and value it.

The critical point is that before we can change our European, utilitarian paradigm of the Pinchot era, which "forces" us to view the forest and all it contains simply as commodities to be endlessly exploited, we must devise a new paradigm, one that recognizes and accepts the difference between — and the scientific validity of — the biological capital that drives the health of the forest as opposed to economic capital into which we humans convert the products of the forest to fulfill our desire for monetary gain. In this new paradigm, we must also view the forest as a living organism with which we cooperate, and through such cooperation we are allowed to harvest products as the ecological capability of the forest permits over the long run — this is sustainability.

Indigenous Forests and Ecological Sustainability

In our burgeoning, product-oriented society, one of the most insidious dangers to indigenous forests — that which has experienced minimal disruptive human intrusion — be they old or young, is the sadly mistaken perception that there is no value in maintaining such a forest for its potential. By potential, we mean its value as a blueprint or what an ecologically sustainable forest is, how it functions, its educational value, its spiritual value, or any other value that does not turn an immediate, visible, economic profit.

This short-sightedness is understandable considering that, as previously stated, (1) the world view within the current paradigm of mechanical reductionism and continual economic expansion is one of economic conversion potential, which means that nothing of Nature has a value unless it can be converted into something else — trees into boards; (2) we in Western industrialized society focus predominantly on utilizable products from the ecosystem rather than on the processes that produce the products; (3) renewable natural resources are largely manifested and thus managed for aboveground, whereas many of the processes that produce them are hidden belowground and hence ignored; (4) we still live in an economic world very much driven by each person looking out first and foremost for himself or herself based on the almost absolute rights of private property; and (5) maximum, short-term, economic gain is the driving force behind our management of renewable natural resources and our society.

When these points are taken together, they form the foundation of Western industrialized economic culture. Reared with this historical background, most people find it difficult to really understand the risks to society's future that accompany the current violations of remaining "natural" areas either in principle or in fact. Consider that, in addition to representing a collection of indigenous species of both plants and animals with a given amount of genetic diversity, each protected area of indigenous forest, whether old or young, also represents a repository with a portion of the world's healthiest ecological processes and their attendant functions.

Part of the process of maintaining ecological resilience in the landscape is setting aside an ecologically adequate system of natural areas of indigenous forest, including commercially available old growth, as an unconditional gift of potential knowledge for the future. In so doing, present and future generations have repositories not only of species, which more often than not are region-specific, but also of processes, which more often than not are worldwide in principle and application.

From such repositories, in addition to monitoring human-caused changes and maintaining habitat for particular species, it will be possible to learn how to maintain, restore, and sustain biological processes in various portions of the ecosystem. In this sense, reserves of indigenous forest, including commercially available old growth, are the parts catalog and the maintenance manual not only for forests of the present but also for forests of the future.

Here we need to pause and carefully consider what British historian Arnold Toynbee learned when he asked the critical question of why 26 great civilizations fell. He concluded that they collapsed because they could not or would not change their direction, their way of thinking, to meet the changing conditions of life. "We cannot say [with certainty what will happen]," wrote Toynbee, "since we cannot foretell the future. We can only see that something which has actually happened once, in another episode of history, must at least be one of the possibilities that lie ahead of us."

For us, the adults of the world, to grasp this lesson in terms of the children of today and of tomorrow and beyond, we must ask ourselves if society can continue to afford the environmental costs of the economics of extinction. We must ask if we have become so materialistically myopic that we are willing to risk losing the ability to have sustainable forests by pursuing the short-sighted, short-term, economic windfall to be had by liquidating the remaining indigenous forests, including the remaining commercially available old growth.

In this context, it is imperative to understand that contemporary humanity has not "*re*forested" a single acre, because no one has as yet planted and grown a biologically sustainable forest on purpose, which is not to say that it cannot be done — because it can. What we and the rest of the world have done and are doing under the guise of "forestry" is at best trading our forests in on simplistic monocultural tree farms and at worst deforesting the rest of the world through such practices as clear-cut logging, salvage logging, slash-and-burn agriculture, and the acceptance of uncontrolled urban sprawl.

In any event, forests and monocultural tree farms are not ecologically synonymous, even under the best of circumstances. If you are skeptical about the validity of this statement, as many foresters seem to be, you might ask what role indigenous forests, especially old-growth forests, play in the notion of biological sustainability. This said, we recognize that tree farms, as the initial stage, can facilitate the positive reversal of denuded land back toward a healthy, forested ecosystem.

Old-Growth Forests and Biological Sustainability

There are many valid reasons to save the remaining commercially available old-growth forests from liquidation and extinction, as many perhaps as there are for saving tropical forests. One is that old-growth forests, such as those of the Pacific Northwest, are both beautiful and unique in the world. Another is that the ancient trees are among the oldest living beings on Earth and as such not only inspire spiritual renewal in many people but also are among the rapidly dwindling living monarchs of the world's forests. They are unique, irreplaceable, and finite in number. And they exist precisely once in forever!

We can perhaps grow large trees over two to three centuries, but no one in the U.S. has yet done that on purpose. If they do, such trees will not be Nature's trees; they will be humanity's trees. And while they may be just as beautiful as those created by Nature, they will be different in the human imagination. Be that as it may, it is doubtful at best that the timber industry would commit to such an endeavor because of the economic costs and biological uncertainties involved.

A third reason to save the remaining commercially available old-growth forests is that a number of organisms, such as the northern spotted owl, the northern flying squirrel, and the red-cockaded woodpecker, either find their optimum habitat in these ancient forests or require the structures provided by them, such as the large live trees, the large declining trees, the large snags, and/or the large fallen trees. A fourth reason is that each old-growth tree is a "carbon sink," a storehouse of immobilized carbon, the storage of which reduces the carbon dioxide in the atmosphere and thereby has a positive influence on the greenhouse effect.[24]

And a fifth reason is that old-growth forests are the only living laboratories through which we and the future may be able to learn how to create sustainable forests — something no one in the world has thus far accomplished.

As a living laboratory, old-growth forests serve four vital functions: first, they are our link to the past, to the historical forest. The historical view tells us what the present is built on, and together the past and the present tell us what the future may be projected on. Because the whole forest cannot be seen without taking long views both into the past and into the future, to lose the remaining old-growth forests is to cast ourselves and our children adrift on a sea of almost total uncertainty with respect to the creation and sustainability of forests of the future.

To have sustainable forests, we need to be able to know, to learn, and to predict. We must therefore remember that knowledge is only in past tense; learning is only in present tense, and prediction is only in future tense. Without significant amounts of old-growth forests, which are allowed to function as much as ecologically possible in the absence of direct, human intervention, we eliminate learning, curtail our knowledge, and greatly diminish our already limited ability to predict.

Second, we did not design the forest, so we do not have a blueprint, a parts catalog, or a maintenance manual with which to understand and repair it. Nor do we have a service department in which the necessary repairs can be made. Therefore, how can we afford to liquidate the remaining old-growth forests, which act as a blueprint, parts catalog, maintenance manual, and service department? They are our only hope of understanding the sustainability of both the potential forests of the future and, as much as possible, our simplified economic tree farms.

Third, we are playing "genetic roulette" with our economic tree farms, both present and future. What if our genetic engineering, our genetic cloning, our genetic streamlining, and our genetic simplifications run amuck, as they so often have around the world? Indigenous forests, be the old or young, are thus imperative, because they — and only they — contain the entire genetic code for living, healthy, adaptable plants and animals that in relationship comprise the forest.[7]

Fourth, intact segments of the old-growth forest, including those that are commercially available, from which we can learn will allow us to make the necessary adjustments in both our thinking and our subsequent course of actions to help assure the sustainability of forests manipulated by humanity. If we choose not to deal with the heart of the issue of old-growth forests, we will find that reality is more subtle than our understanding of it and that our "good intentions" will likely give bad results over time.

There are many valid reasons to save the remaining old-growth forests, but there is only one reason for liquidating them — short-term economics. Economics, however, is the common language of industrialized society. Is it not wise, therefore, to carefully consider whether a necessary part of the equation for maintaining a solvent forest industry is to save what little old-growth forest remains?

Cutting the remaining commercially available old-growth forests will serve only a small proportion of the immediate generation of humans for only a few years, whereas protecting these same forests will serve all of society within all generations to come. We must therefore be exceedingly cautious lest faulty economic reasoning isolates us from the evidence that, without ecologically sustainable forests, we will not have an economically sustainable timber industry, and without an economically sustainable timber industry, there will continue to be a growing number of human communities without a sustainable economy.

Thus, if we liquidate the remaining commercially available old-growth forests — our living laboratories — and our economic tree farms fail, as tree farms are failing over much of the world, there will be no forest industry, and we will have further impoverished our souls and those of the generations of the future through our myopic drive for maximum, short-term profits. American writer Minna Antrim put it well when she said that "experience is a good teacher, but she sends in terrific bills." Unless our minds and our hearts are set on maintaining an ecologically sustainable forest, each succeeding generation will have less than the preceding one, and their choices for social survival will be equally diminished.

PART TWO: SUSTAINABLE FORESTRY THROUGH THE PROCESS OF CERTIFICATION

We belong to the ground.
It is our power,
And we must stay close to it
or maybe we will get lost.

— Narritjin Maymuru Yirrkala,
an Australian aborigine

Chapter 2

Certification and Trusteeship

Before certification and trusteeship can be fruitfully discussed, you must be able to understand and integrate two perspectives of time, that of a clock and that of an hourglass. Time as measured by the ticking of a clock is constant in tempo. With a clock, one sees the hands move from second to second, minute to minute, and hour to hour — as round and round the clock's face they go. To a child, however, time seems to drag, even stand still; to an older person, time seems to fly, despite the fact that watching a clock's hands make their appointed rounds belies both the impatience of youth and the sensation that time is fleeting in old age.

If, on the other hand, one observes an hourglass as a measure of time, one has the distinct impression that time is running out, like the sand pouring at the beck and call of gravity from the top of the hourglass through the small central hole in the middle of the hourglass to the bottom. Most adults view time with a growing sense that theirs is running out, so they must grab all of life they can before their time is spent, which usually equates to the acquisition of materialism in the known "safety" of the status quo. It is the sense of impending loss as time "runs out" that causes people to avoid change, which they fear will have emotional suffering woven into it because of this impending sense of loss.

In reality, of course, time does not run out; our bodies expire instead. And it is precisely the duel sense of time running out and the demise of our bodies that causes many people to seek a way to continue their sense of being in the world, like the continual ticking of the clock. One way to accomplish such continuance is through a living trust.

If we have the courage and the willingness to adopt and implement the concept of "biologically sustainable forestry," which in fact means forests and not economic tree farms, then the notion of ever-adjusting relationships — adaptive management — becomes the creative energy that guides a vibrant, adaptable, ever-renewing profession through the present toward the future. And because biologically sustainable forestry honors the integrity of both society (intellectually, spiritually, and economically) and its environment, the outcome fits into the concept of a "biological living trust" in which a system's function defines the system. That is to say that the function defines the necessary composition of the system, which in turn defines the structure, and it is therefore by its function that we must learn to characterize a system.

A biological living trust is predicated on "holism" in which reality consists of organic and unified wholes that are greater than the simple sum of their parts. The following are basic assumptions on which a biological living trust is founded:

- Everything, including humans and nonhumans, is an interactive and interdependent part of a systemic whole.
- Although parts within a living system differ in structure, their functions within the system are complementary and benefit the system as a whole.
- The whole is greater than the sum of its parts because how a system functions is a measure of its ecological integrity and biological sustainability.
- The ecological integrity and biological sustainability of the system are the necessary measures of its economic health and stability.
- The biological integrity of processes has primacy over the economic valuation of components.
- The integrity of the environment and its biological processes have primacy over human desires when such desires would destroy the system's integrity (= productivity) for future generations.
- Nature determines the necessary limitations of human endeavors.
- The relevancy of knowledge depends on its context.
- The disenfranchised as well as future generations have rights that must be accounted for in present decisions and actions.
- Nonmonetary relationships have value.

In a biological living trust, the behavior of a system depends on how individual parts interact as functional components of the whole — not on what each isolated part is doing — because the whole is understood through the relation/interaction of its parts. Thus, to understand a system we need to understand how it fits into the larger system of which it is a part, which gives us a view of systems supporting systems supporting systems, *ad infinitum*. We therefore move from the primacy of the parts to the primacy of the whole, from insistence on absolute knowledge as truth to relatively coherent interpretations of constantly changing knowledge, from an isolated personal self to self in community, and from attempting to solve old problems with old thinking to creating new concepts tailored specifically to today's changing social–environmental context.

In a biological living trust, individual people — as well as their relationships among one another, Nature, and their communities — have value and are valued, as are all living beings. To survive, therefore, forestry must advance beyond resisting change as a condition to be avoided (clinging to the current, linear, reductionistic mechanical world view) and embrace change as a process filled with hidden, viable ecological-social-economic opportunities in the present for the present and the future.

On the other hand, a living trust in the legal sense is a present transfer of property, including legal title, into trust, whether real property or personal property, livestock, interests in business, or other property rights. The person who creates the trust can watch it in operation, determine whether it fully satisfies his or her expectations, and, if not, revoke or amend it.

A living trust also allows for delegating administration of the trust to a professional trustee, which is desirable for those who wish to divest themselves of managerial responsibilities. The person or persons who ultimately benefit from the trust are the beneficiaries. Can a forest be such a living trust in the legal sense?

A Certified Forest as a Living Trust

All we adults have to offer our children and the generations of the future are options (which are choices to be made) and some things of value from which to choose. Those options and things of value, both biological and legal, can be held within the forest as a living trust, of which we adults are the legal caretakers or trustees for the next generation (who are the beneficiaries). Although the concept of a trustee or a trusteeship seems fairly simple, the concept of a trust is more complex because it embodies more than one connotation; consider, for example, a legal living trust.

The forest is a "living trust" in the present for the future. A living trust, whether in the sense of a legal document or a living entity entrusted to the present for the future, represents a dynamic process. Human beings inherited the original living trust — the living world of which a forest is but a part — before legal documents were even invented. The Earth as a living organism is the living trust of which we are the trustees and for which we are all responsible, whether on public lands or private lands.

Public Lands

Throughout history, administration of our responsibility for the Earth as a living trust has been progressively delegated to professional trustees in the form of elected or appointed officials. In so doing, we empower them with our trust (another connotation of the word, which means we have firm reliance, belief, or faith in the integrity, ability, and character of the official who is being empowered).

On public lands, such empowerment carries with it certain ethical mandates, which in themselves are the seeds of the trust in all of its senses, legal, living, and personal:

1. "We the people," present and future, are the beneficiaries and the elected or appointed officials are the trustees.
2. We have entrusted our officials to follow both the letter *and the spirit* of the law in the highest sense possible. In speaking of the spirit of the law, the words of Helen Keller beckon to be heard: "I long to accomplish a great and noble task; but it is my chief duty to accomplish humble tasks as though they were great and noble." Such is the life of a public servant.
3. We have entrusted the care of public lands (those owned by all of us), whether forested or otherwise, to officials through professional planners, foresters, and other people with a variety of expertise, all of whom have sworn to accept and uphold their responsibilities and to act as professional trustees in our behalf.
4. We have entrusted to these officials and professionals the living, healthy forest. Through the care of these folks, it is to remain living, healthy, and capable of benefiting both present and future generations. At this juncture, behavioral psychologist Abraham Maslow might have pointed out to these officials and professionals that the healthiest people are those dedicated to something greater than themselves and that, by transcending themselves in the service of others, they are best able to express their own highest potential.
5. Because we entrusted public lands as "present transfers" in the legal sense, we have the right to either revoke or amend the trust (the empowerment) if the trustees do not fulfill their mandates.
6. To revoke or amend the empowerment of our delegated trustees if they do not fulfill their mandates is both our legal right and our

moral obligation as individual, hereditary trustees of the Earth, a trusteeship from which we cannot divorce ourselves.

How might this work if we are both beneficiaries of the past and trustees for the future? To answer this question, we must first assume that the administering agency is both functional and responsible, be the public lands in federal, state, county, or municipal ownership. The ultimate mandate would then be to pass forward as many of the existing options (the capital of the trust) as possible.

These options would be forwarded to the next planning and implementation team (in which each individual is a beneficiary who becomes a trustee) to protect and pass forward in turn to yet the next planning and implementation team (the beneficiaries that become the trustees) and so on. In this way, the maximum array of biologically and culturally sustainable options could be passed forward in perpetuity.

If, however, the officials and/or professionals did not fulfill their obligations as trustees to our satisfaction, then their behavior could be critiqued through the judicial system, assuming that the judicial system is both functional and responsible. Thus the carefully considered effects embodied in our decisions as trustees of today could create a brighter vision for the generations to come, who are the beneficiaries of the future when they stand in their today. In order for this to happen, however, we must first mandate that the administering agency and the judicial system be made both functional and responsible, something we have seldom chosen to do completely. Now, how might this work on private lands?

Private Lands

Instead of delegating to professional trustees the admistration of a private forest as a living trust, the landowner empowers himself or herself with that task. Such empowerment also carries with it certain ethical mandates, which are overseen by local, state, and, at times, federal regulations. Since no single individual can know all things, a landowner might hire a professional consultant to counsel and/or take care of the property or apply to have the forest certified as sustainable.

If the latter course is chosen, a carefully selected team of individuals with a variety of expertise examines the land and tells the owner in general terms what needs to be done to meet the requirements for certification. If, after reviewing the report, called a scoping, the landowner wants to proceed, a full assessment team arrives and counsels the owner on what must be done to meet the standards for certification of the forestland as biologically sustainable, as will be discussed in detail later in this book. In this case:

1. The landowner becomes the trustee, and whoever inherits or purchases the land is the primary beneficiary. Everyone living within the landscape, however, benefits to some extent as the acres of forest are rendered as sustainable as possible, which increases not only the long-term productivity of the individual acres but also ensures a greater measure of Nature's free services, such as clean air, for all inhabitants of the immediate landscape who become the peripheral beneficiaries.

2. The landowner has, through the certification process, committed himself or herself to follow both the letter *and the spirit* of the law of a biological living trust in the highest sense possible, which calls to mind an astute observation by French author Antoine de Saint-Exupery: "It is only with the heart that one can see rightly; what is essential is invisible to the eye."

3. The landowner has entrusted a team of people with varied expertise, through the certification process, to help him or her determine what needs to be done to care for the land in the most ecologically sustainable manner as the trustee of a biological living trust on behalf of the future owner.

4. The landowner entrusts the certification team to help him or her ensure the health of the forest as a biological living trust in such a way that it is capable of benefiting both those who depend on it in the present and those who will depend on it in the future.

5. Because the landowner entrusts his or her land as a "present transfer" in the biological sense to the certification process, the certifier (acting as a joint trustee), in our case SmartWood, has the right to either revoke or amend the trust if the owner does not fulfill his or her acknowledged agreements.

6. If, however, the certifier (acting as a joint trustee) does not fulfill its acknowledged obligations to the satisfaction of the land-owner, the landowner could refer the case to the Forest Stewardship Council, which oversees the certifiers (as will be discussed later), and/or choose another certifier, something that, to our knowledge, has not happened.

Thus the carefully considered effects embodied in a landowner's decisions as a trustee of today could create a brighter vision for the next person(s) who owns the land. This is all well and good, but why is the emphasis on sustainable forestry and trusteeship so important with respect to the scattered, small acreages of the average landowner?

The Importance of Certification

Although we have eluded to certification of forestlands as a critical pathway through time and space by which parents (trustees) can pass to their children, grandchildren, and beyond (beneficiaries) the gift of choices and some things of value from which to choose, we need to examine further the reasons we say this, reasons that are both ecological and social, beginning with personal freedom.

Freedom

There is no such thing as complete freedom, either in Nature or in society, because everything is defined by its relationship to everything else. Hence, there is no such thing as an independent variable in any interactive system. This being the case, every landowner's actions affect the ecological integrity not only of the acres he or she owns but also of the landscape as a whole for better or ill in both time and space. Consequently, every landowner's actions also affect all people within the water catchment in which he or she lives, the drainage basin into which the water catchment flows, and ultimately the ocean into which the drainage basin empties.

Certification is designed to balance the limitations of personal freedom with the enhancement of collective freedom. This means that the purpose of certification is to help assure that the affect of a landowner's decisions and subsequent actions on his or her private parcel of land will be ecologically positive to the greatest extent possible because each acre is an individual part of a mosaic that in the collective forms the patterns of vegetation one sees across a given landscape. Such patterns are in turn imperative to the ecological integrity of a landscape because it is the self-reinforcing feedback loops of hidden biophysical interrelationships among plants and animals that confer on the landscape a degree of stability and thus a collective freedom over time for all people who live there.

Patterns across a landscape are also a matter of scale. A journey of a thousand miles, says an old Chinese proverb, begins with a single step; likewise, the sustainability of an entire landscape begins with the sustainability of a single acre. This proverb fits well with a statement once uttered by an anonymous person who mused that "great opportunities to help others seldom come, but small ones surround us every day." For the trustee of forestland, each acre is such a small opportunity to serve all of humanity.

If, for instance, a landowner wants to have his or her forest certified, for whatever personal reason, that is a beginning. From then on, each landowner who earns the certification of his or her land becomes not only the trustee of his or her own biological living trust but also a member of a larger trusteeship, that of the landscape as a biological living trust, which is the collective of each person who earns certification of his or her own acreage. Consider further that to hold a landscape of varied ownerships in a living trust requires a multitude of owners who are committed to earning certification for their respective acreages.

To this end, a study published in the 1998 *Journal of Forestry* has found, through a survey of landowners, that they are already predisposed to the basic principles of ecosystem management. Landowners in western Massachusetts, for example, already realize that their actions cross property lines and affect land elsewhere. But while they value the components of their landscape and want to leave a healthy forest as their legacy to the next generation, they need to see how their private parcels fit into the bigger picture before they can work together to care for their properties as an ecosystem across a landscape.[25]

The survey also found that landowners place a high priority on privacy, which raises the question of how this sentiment might affect a landowner's willingness to engage neighbors in a cooperative, coordinated action, which is a prerequisite for a sustainable landscape. This priority on privacy raises another question; namely, what trade-offs are owners willing to make among current and future uses of their respective lands as they affect landscape-level impacts?[25]

Most private, nonindustrial forest owners seek certification out of a serious commitment to the well-being of their forests, according to a recent study in the *Journal of Forestry*.[26] In so doing, they consider the cost of meeting contractual conditions and agreements as part of the normal business of improving their forestry practices. Although higher profits for their timber have not always been realized, many feel that the process of certification allows them to examine their practices without getting embroiled in a political debate that a discussion of forestry often triggers.

The empirical findings of the study are presented as three conclusions:

1. Owners of small forestland operations tend to seek the certification process as a way of satisfying their intrinsic needs to learn, to feel personally good about what they are doing, to fulfill their sense of obligation to society as a responsible member, and to get a reality check of their accomplishments through external validation. In contrast, managers of industrial and public forestlands tend to seek the certification process as a way of satisfying extrinsic demands, such as increasing profits, defending their access to and share of the market, and protecting their "social license" to continue their operations through external validation of their management as ecologically sound and human friendly.

2. The main barrier to seeking certification by small enterprises is likely to be the high direct costs, such as paying for the certification process. Beyond that, indirect costs, such as the possible necessity of upgrading roads, may also be high and make implementation of the requirements for certification economically difficult. Be that as it may, landowners who entertain the notion of certification already perceive themselves to be practicing sustainable forestry and thus do not foresee large indirect costs. But when faced with such costs, landowners generally accept them as expected costs of doing business, irrespective of the requirements for certification.

3. Once engaged in the certification process, landowners modify their visions, goals, and objectives and in so doing find new benefits that reinforce their commitment to the certification process. Even when economic rationale is the initial motivation for seeking certification, participation in the process highlights the benefits of continued personal growth through learning and improving one's forestry practices.

Despite some frustration with the indirect economic costs of certification, no one in the study planned to quit. Most people emphasized the noneconomic benefits, while others adopted a long-term perspective of waiting patiently for the development of differentiated markets for certified forest products. In addition, several participants in the study saw an advantage in the reflection and innovation that the process of certification brought to their forestry practices. Such reflection is likely to result in more natural regeneration; more attention to the collection of local seeds and seed zoning, which necessitates better seed certification; and maintenance of greater diversity, both biological and genetic.[27]

With respect to public forestlands, managers in Pennsylvania and Minnesota have concluded that, while certification still needs adjustments and may not be for everyone, it has a place as a voluntary, market-driven program that both recognizes and fosters better forest management wherein the benefits outweigh the perceived risks and costs. They also concluded that certification under the Forest Stewardship Council is compatible with, and in some cases complementary to, other approaches to assessing ecological sustainability in forest practices, such as the Sustainable Forestry Initiative.[28]

Thus, every landowner who earns certification adds to the critical mass necessary to repair and maintain a landscape in a sustainable condition, despite the fact that no single ownership will or can be totally sustainable. In this way, as in all ways, the actions of a single individual affect the social circumstances of all individuals because it costs less in biological, economic, and social capital to maintain a healthy, sustainable landscape than it does to repair one, especially since the cost of repairs is inevitably compounded exponentially the farther into the future they are projected. All this, however, hinges on one's sense of personal values.

A Sense of Values

An individual's personal feelings and values cannot be legislated from the top down; they can only be planted, nurtured, and grown from the bottom up, one "individual" at a time. This is an important concept because a democracy, such as ours in the U.S., is by nature restrictive in that laws are passed to protect the rights of the majority from the transgressions of the minority, and herein lies the problem. Everyone pays the same price for a restrictive law by losing a heretofore legal option, even if one has never abused anyone else's legal rights.

It is precisely because of the ubiquitous limiting nature of legal restrictions to personal and social freedom that certification is so important. Certification is voluntary compliance with the highest standards of the democratic principles embodied in acting like a psychologically mature adult concerned with the welfare of one's neighbors — both in physical space (as in the person living next door) and in time (as in the children who are the next generation). By doing what is ecologically best for one's land, a landowner is passing foreward a personal legacy of the highest moral character and in so doing validates the nobility of one's trusteeship. If every landowner were to act accordingly, legally imposed regulations would be unnecessary. Impossible, one might say, because it goes against human nature to be altruistic. Why bother, considering that a person is but a single individual going against both the tide of sanctified rights of private property and sacred consumerism in American society. In answer to this question, think of the cumulative effect of snowflakes in a winter storm.

To understand the value and power of each person in the context of his or her collective thoughts and actions on his or her own acres of private forestland, pretend for a moment that we humans are snowflakes. We are part of the first snow of winter. At first we fall in small numbers, perhaps 1, 2, 10, or even 20 at a time, and we are hardly noticeable in the vastness of open space. One by one we fall out of a quiet sky, as we whirl and spin to Earth, and in our falling, we magnify one another.

The pioneers were the first flakes to fall (the first people to earn certification of their forests); they landed on warm soil and melt, disappearing without an apparent trace. But are they really lost? Have they really had no effect? Each flake that lands on the soil, only to melt and disappear, gives its coolness to the soil until, after enough flakes have landed and melted, the temperature of the soil drops (which is analogous to the initial critical mass of people who have earned certification of their forestland and begin to visibly affect the sustainability of the landscape).

Finally, because of the cumulative effect of all the flakes (all the certified acres) that have gone before, the soil has cooled enough for us, you and me, to survive as we land, and still the flakes fall, now in the millions, each individual in its own time and in its own way. It snows all night, and by morning, a glittering, transformed world greets the rising sun. As far as the eye can see is a world of winter white, one flake at a time, as we add our collective beauty to the wonder of the Universe (until the entire landscape is sustainable).

For certification to be accepted, however, it must have certain characteristics that both create and foster trust among people.[29] Forest certification must:

- Be credible to consumers and nongovernment conservation organizations
- Develop measurable criteria that are as objective as humanly possible
- Perform each assessment in a reliable and independent manner
- Be independent from parties with vested interests
- Be cost-effective
- Be philosophically and functionally transparent to allow external critique
- Be institutionally and politically adapted to local conditions
- Be vision and goal oriented and effective in achieving reachable objectives
- Be acceptable to all parties involved
- Use regional-level forestry criteria that are compatible with generally accepted principles of both ecology and international forest certification

In short, forest certification can best be understood as an instrument of policy that is intended to foster ecologically sound forestry by providing market incentives. Certification can also be a powerful tool for measuring a landowner's progress toward achieving his or her important but ever-elusive goal of sustainable forest management. In this way, certification not only helps to make landscapes biologically sustainable but also helps people to sort out those questions that rightly belong to science and those that rightly belong to social values.

Questions of Science vs. Questions of Social Value

The success of forest certification depends on our ability to see the forest as a functional whole; by this we mean the ability to see the commercial forest products in terms of a biologically sustainable system, as well as a biologically sustainable system in terms of its products. Unfortunately, we usually juxtapose product points of view and systemic points of view into a category of *either* products or forest health — instead of products *and* forest health — when asking questions of value. We then try to force science to answer the questions. But science, which theoretically is the free pursuit of biophysical knowledge for its own sake, is the language of the intellect and is concerned with how and why the universe works as it does. Scientific inquiry is not designed to deal with values, which are the language of the heart.

While both languages are necessary if human culture and its manifold environments are to be mutually sustainable, our intellectual illusion of definitive knowledge is the salient point, not the state of our ignorance. And it is exactly because we are so certain of our knowledge that we are often so abysmally unaware of our ignorance.

In our experience, the more product oriented a person is, the more certain he or she is of his or her knowledge and the more resistant he or she is to change, seeing it as a condition to be avoided because he or she feels a greater sense of security in the known elements of the status quo, especially where money and private property are concerned. The greater the product orientation, the more black-and-white one's thinking tends to be, which may have led American psychologist William James to observe that "a great many people think they are thinking when they are merely rearranging their prejudices."

On the other hand, as Helen Keller once said, "Security is mostly a superstition. It does not exist in Nature. Life is either a daring adventure or nothing." The more of a systems thinker a person is the more likely he or she is not only to agree with Helen Keller but also to risk change by thinking outside of the socially negotiated box on the strength of perceived unseen possibilities.

Our experience indicates that people who seek to earn certification for their forestland tend to be systems thinkers. As such, they have a better grasp of the difference between those questions that belong in the realm of science and those that belong in the realm of social values. A systems thinker is therefore more likely to be concerned with the welfare of others, including those of the future and their nonhuman counterparts. Systems thinkers also tend to be concerned with the health and welfare of planet Earth in the present for the future. And systems thinkers more readily accept shades of gray in their thinking than do product-oriented people.

Product-oriented people tend to focus on individual pieces of a system, its perceived products, in isolation of the system itself, whereas systems thinkers tend more toward a process approach to thinking. A person oriented to seeing only the economically desirable pieces of a system seldom accepts that removing a perceived desirable piece through over-exploitation or an undesirable piece through purposeful extirpation can or will negatively affect the system's productive capacity as a whole.

In contrast, a systems thinker sees the whole in each piece and is therefore concerned about ignorant tinkering with the pieces because he or she knows such tinkering may inadvertently upset the desirable function of the entire system. A systems thinker is also likely to see himself or herself as an inseparable part of the system, whereas a product-oriented thinker normally has himself or herself set apart from and above the system. And a system thinker is willing to focus on transcending the issue in whatever way is necessary to frame a vision that protects the system itself for the good of the majority in both present and future generations.

Product-oriented thinking argues to retain the old reductionistic mechanical world view as its premise for decision-making. Systems thinking argues for an evolving unified world view even though it is only now emerging into our consciousness as a paradigm founded on the notion of sustainability.

The conflict in decision-making, therefore, is between product-oriented and systems-oriented (process-oriented) values based on different world views, something science can approach only indirectly, politicians often studiously avoid, and those who seek to earn certification must forthrightly address. Certification may therefore be the precursor to placing sound ecological limitations on the unfettered rights of private property in a voluntary way that gives voice to the welfare of future generations. The notion of human welfare, a critical component of which is sustainable forestry through forest certification, brings us to a discussion of the gross domestic product, eco-efficiency, and genuine progress indicator as measures of sustainability.

Gross Domestic Product, Eco-Efficiency, and Genuine Progress Indicator

According to the reductionist mechanical world view, which is today overlain with the notion of continual economic expansion, the economic process of producing and consuming material goods and services has no deleterious effects on the ecosystem because the assumptions are that natural resources are limitless and any unintended effects of the economic process, such as pollution and environmental degradation, are inconsequential. In contrast to the dominant world view, however, the paradigm of sustainability is neither mechanical nor reversible; it is entropic, which means that the Earth's resources and its ability to absorb and cleanse the waste produced by humanity's economic activities are both finite.[19]

By tying the economic process to the entropy of the physical world, as mentioned earlier, economist Georgescu-Roegen pointed out that for Western industrialized society to survive with any semblance of dignity, there must be a shift from the old reductionist mechanical world view to a paradigm built around sustainability. In making that point, however, he posed the unspoken question of how one measures sustainability in terms of human welfare. For the sake of discussion, three potential measures will be considered: Gross Domestic Product, Eco-efficiency, and Genuine Progress Indicator.

Gross Domestic Product

The Gross Domestic Product, which is nothing more than a measure of total output (the dollar value of finished goods and services), tells very little in and of itself because it assumes that everything produced is by definition "goods," including people.[19] William Bennett, who was President Reagan's Secretary of Education, observed that "socialism treats people as a cog in a machine of the state; capitalism tends to treat people as commodities." As such, the Gross Domestic Product is an intellectual measure of the size of the U.S. economy, the amount of money that exchanges hands in a strictly additive sense, like an adding machine that cannot subtract, and thus makes no distinction between benefits and costs (credits and debits), productive and destructive activities, or sustainable and nonsustainable activities, in addition to which there is no allowance for the declining quality of human life in the face of environmental degradation.

The reason for this disregard of human welfare is simply that the Gross Domestic Product treats everything that happens in the marketplace as a positive gain for humanity and thereby de facto ignores everything that cannot be converted into money as being unimportant to social well-being, such as the logging practices that destroy habitat for salmon. In this case, both logging and commercial salmon fishing cause money to exchange hands and count as a plus in the valuation of the Gross Domestic Product, even though the degradation of the salmon's habitat caused by logging in the mountains will eventually put the commercial salmon fisher in the ocean out of business. Politicians, however, generally see this decaying quality of human life through a well-worn ideological lens that accepts economic growth as good even as it cannibalizes the family, community, and environment that nurtures and sustains us.

On a more personal note, consider a man dying slowly of cancer who needs three major operations while in the middle of a messy divorce that forces him to sell his home. This man is an asset to the economy from the Gross Domestic Product point of view because he is the cause of so much money exchanging hands.

In the first case, the commercial salmon fisher is faced with a declining way of life because the logging he or she never sees is slowly destroying his or her livelihood. In the second case, the quality of life of the dying man could hardly get much worse.

In both cases, the valuation of the Gross Domestic Product goes up at the unmeasured expense of the commercial salmon fisher who is losing a way of life and the dying man who is losing everything he held dear to forces other than his impending death. This scenario is somewhat analogous to adding (crediting) the amount of each check one writes against one's bank account instead of subtracting (debiting) it.

The significance of this illogical calculation of economic activity revolves around the Gross Domestic Product as the primary indicator of economic growth (the economic score card) from one year to the next in the U.S. As such, when growth in the Gross Domestic Product exceeds three percent, it is usually favorable for incumbent politicians. The danger hidden within the calculation of the Gross Domestic Product as a real measure of economic growth, however, is that it creates a false sense of prosperity and security, especially when growth is rapid because it ignores costs (adding only the benefits) and thus ignores the major problems confronting American society. This is like adding up all of the inflowing cash from cutting and selling one's timber, while ignoring both short-term costs (such as physical wear on roads and equipment and the human labor involved in felling, bucking, and cutting the trees into logs and then yarding, loading, and hauling the logs to market) and long-term costs (such as replacement of machinery and worn-out culverts in roads, the effects of soil compaction, siltation of streams through soil erosion from roads, loss of soil fertility through clear-cut logging, fragmentation of habitats and the consequent restrictions of the Endangered Species Act, and so on).

Money itself as a measure of success is another example of a serious flaw in thinking and valuation where sustainability is concerned because the bottom line in business is always pleaded as the truly important figure.[30] The bottom line, which shows how much profit has been made, is used as a measure of how well a company performs. Too little profit, and a company is deemed inefficient, its management is slack, the full potential of its workforce is not harnessed, its products are out of date, or most damning of all, the company is not competitive in the global economy. Are such damnations true? Is money the only valid measure?

Perhaps a family-owned timber company is making products that are robust and lasting and/or selling its products to people with only a moderate income or those who are somehow disadvantaged. It may be paying its employees higher wages than other timber companies in the belief that all people deserve a living wage. It may be investing heavily in a strategy to protect the ecological integrity of its forestland, or having a noisy mill in a location being increasingly surrounded by people's homes and thus operating only one shift during the week out of respect for the people living in the neighborhood.

In a world where money is the only acceptable measure of success, however, all these considerations count as naught because traditional economists assure us that their linear notion of progress, which means full steam ahead in the strictly material realm, is always the right course of action (ready, **fire**, aim), whereas ecology is a discipline that teaches us the folly of speeding blindly into the future (ready, **aim**, fire). In the scenario of full steam ahead, the quality of the products and the welfare of the people and the environment are all irrelevant in the face of a bottom line that is not performing as desired. The irony is that the bottom line actually accounts for the last 10 percent of income, the part that constitutes the profit after the other 90 percent has been paid on the way to the 10 percent, and yet the last 10 percent overrules and thus over-shadows the 90 percent.

This type of valuation clearly points out that market economics places value on that which is scarce (as profit is considered to be) instead of the real work and worth of people, that is, being a loving, caring, honest, just, and thoughtful person and neighbor. If we are to keep the softer social capital of mutual caring from becoming scarce, we must reward it. This, however, is one of the many areas in which the last 10 percent of the dollars, squeezed into profit margins at the expense of the 90 percent along the way, is simply not effective in meeting human welfare because "they" do not build families or communities or tackle poverty or protect the environment.

Clearly, therefore, the Gross Domestic Product, with its myopic focus on dollars and its flawed logic, cannot be a measure of sustainability as it relates to human welfare. If not Gross Domestic Product, then what could speak for human welfare? Many industrial participants of the 1992 Earth Summit in Rio de Janeiro, Brazil, touted a strategy of "eco-efficiency" that would not only refit the machines of industry with cleaner, faster, and quieter engines but also allow unobstructed prosperity while simultaneously protecting both economic and corporate structures.

Eco-Efficiency

Industrialists hoped that eco-efficiency would transform the economic process from one that takes, makes, and wastes into a system that integrates economic, environmental, and ethical concerns. Here you might ask what this notion of eco-efficiency is that industrialists around the world herald as their chosen strategy for change.[31]

Eco-efficiency is a term that primarily means doing more with less, a precept that Henry Ford was adamant about when he wrote in 1926, "you must get the most out of the power, out of the material, and out of the time." His lean and clean operating policies saved his company money by recycling and reusing materials, reducing the use of natural resources, minimizing packaging, and setting new standards of human labor with his timesaving assembly line.

Although eco-efficiency is a well-intentioned concept that looks good on the surface, it is still within the bowels of the reductionist mechanical world view with its current overlay of economic expansionism, and thus is little more than an illusion of change. Rather than focusing on a new way of thinking, such as how to *effectively* save the environment, industrialists once again attached their hope to efficiency — the swan song of the environment — with which, unconsciously perhaps, they have set themselves up to quietly, persistently, and completely commercialize the entire world. This is but saying that eco-efficiency, while it aspires to make the old world view less destructive, languishes from the fatal flaws hidden within the embrace of such destructive practices in the first place.

To view the fatal flaws inherent in the tenets of eco-efficiency, we will design forestry as a retroactive system that not only allows but also encourages people to spend the inherited forests of the world as though there were no tomorrow and to pass the bill forward to the generations of the future. Such a system would function as follows:

- Annually clear-cut as much timber, primarily old growth, as one could sell, preferably as whole logs, overseas.
- Measure prosperity by economic activity and success by the automation that eliminates people's jobs while increasing the profit margin.
- Measure progress as a continual technological advancement in the utilization of wood fiber from ever-younger trees.
- Promote personal self-interest, which requires thousands of complex and often competing regulations to keep self-centered, greedy people from clear-cutting entire landscapes.
- Encourage clear-cutting the entire riparian zone right down into the stream bottom.

- Leave nothing as a reinvestment of biological capital in the soil.
- Erode and ultimately destroy biological, genetic, and functional diversity through centralized corporate economic competition that converts as much of the world's forests as humanly possible into quick monetary profits.

If the above system were refitted with the current notion of eco-efficiency, it would look something like this:

- Annually clear-cuts *fewer* acres and purposefully *hides* them
- Measures prosperity by *less* economic activity and success by introducing automation *more slowly*
- Promotes *less blatant* personal self-interest by *meeting or exceeding* many or most of the complex and often competing regulations
- *Encourages* saving a minimal buffer zone of nonmerchantable trees but only along streams with anadromous fish
- *Encourages* leaving two nonmerchantable logs per acre as a reinvestment of biological capital in the soil
- *Standardizes and homogenizes* biological, genetic, and functional diversity by replacing forests with cloned fiber farms for corporate economic benefit

Clearly, while eco-efficiency aspires to make the reductionist mechanical world view more benign through reduction, reuse, and recycling, it does not stop these economically driven processes of exploitation, needless overproduction, acquisitiveness, and pollution. The real message of eco-efficiency is to restrict industry and slow or curtail growth — to put limitations on the creative and productive capacity of humankind. This message is simplistic, however, because Nature itself is highly industrious, creative through unpredictable novelty, astonishingly productive, and even "wasteful" when viewed in the short term. The salient point is that Nature, unlike human industry, is *effective* — not efficient.

Consider the pine, which annually casts billions of pollen grains to the vagaries of the ever-shifting wind so that a few might land in just the right place at just the right time to consummate the union of male and female gametes to form a few viable seeds. The seeds, in turn, must then ripen and drop to the soil in a place conducive to their germination and growth, all the while beset by the unpredictable elements of weather and the potential for a vast array of hungry microbes, fungi, insects, birds, and mammals to find and eat them — all of this so that a few, a very few, new pines might germinate in sufficient numbers to replace those that died and thus maintain the species. There is little, if any, waste in this apparent inefficiency because pollen and seed alike are sought as food by myriad organisms. Effective, yes; efficient, perhaps not, which brings us to the Genuine Progress Indicator.

Genuine Progress Indicator

The notion of a Genuine Progress Indicator is critical because it is the best (and perhaps the only) accurate way to balance our social values with our growing knowledge of how ecosystems work and the limitations their long-term integrity impose on both the potential and actual sustainability of our activities. Without such an indicator, environmentalists are often viewed and chided as being deviant, radical, subversive, extreme, anti-business, and un-American by business people because environmentalists rate the ecological values embodied in saving old-growth forests and wetlands to be greater than those of economic growth.[32]

On the other hand, environmentalists often view business people as necessarily evil, greedy, and myopic by nature. Most of the problem with the economic point of view espoused by business people, according to Thomas Gladwin, director of the University of Michigan's corporate-environmental management program, is that business executives and managers often lack good cross-training in science, as evidenced by the fact that less than one percent out of 1.2 million articles written by business professors includes the words "pollution," "air," "water," or "energy."[32]

The Genuine Progress Indicator, in contrast to both Gross Domestic Product and Eco-efficiency, is a measure of total economic activity that includes both benefits *and costs* (credits *and debits*).[33] The notion of a Genuine Progress Indicator fits well with the aims of certification, which are to make a forestry operation economically viable so it can (1) pay the workers fairly (providing an economic benefit to the community); (2) reinvest in management, e.g., keep the roads maintained, thin the trees, prescribe burning, etc.; and (3) pay the land taxes — in other words, be able to hand a healthy, economically viable forestland business to the next generation.

In this way, the owner of forestland could measure the true value and ecological well-being of his or her property over time by assigning an economic value to noneconomic indicators, such as the amount of organic material left on a logged site as a biological reinvestment into the soil, the amount of siltation in streams, the temperature of the water in streams, the amount of erosion from roads, and so on. By assigning either a positive or negative value (a credit or debit) to each indicator, the indicators and their respective values can be combined into a single Genuine Progress Indicator for the ecological welfare of a particular owner's forestland.

The Genuine Progress Indicator would then serve not only as a baseline for all subsequent deliberations concerning sustainable forestry but also as a means of measuring the effectiveness of the criteria used in forest certification and subsequent actions based on those criteria. Another function of the Genuine Progress Indicator in forestlands would be to add value to those resources and activities that have no value in terms of the Gross Domestic Product, such as out-sloping roads, saving dead standing trees, protecting a tree with a colony of honeybees in it, and so on. The Genuine Progress Indicator is an important tool because most, if not all, activities in a forest are omitted from valuation within the context of traditional economic measures, which becomes readily apparent when a landowner deliberates over the ecological well-being of his or her forest in terms of traditional economics and as a legacy for the future. Having said this, it is imperative at this juncture to elaborate on some of Nature's services that are omitted from traditional economic valuation.

The inherent services performed by Nature constitute the invisible foundation that is not only the wealth of every human community and its society but also the supporting basis of our economies. In this sense, Nature's services are also the wealth of every owner of forestland. For example, we rely on oceans to supply fish, forests to supply water, wood, and new medicines, rivers to transport the water from its source to a point where we can access it, soil to grow food, forests, grasslands, and so on. Although we base our livelihoods on the expectation that Nature will provide these services indefinitely and free of charge, the economic system to which we commit our unquestioning loyalty either undervalues, discounts, or ignores these services when computing the Gross Domestic Product and the real outcomes of Eco-efficiency. This is but saying that Nature's services, on which we rely for everything concerning the quality of our lives, are measured poorly or not at all.

Because of the importance of Nature's inherent services, usually thought of as ecosystem functions, it is worthwhile to examine one such service in greater detail — pollination. Eighty percent of all cultivated crops (1,330 varieties, including fruits, vegetable, coffee, and tea) are pollinated by wild and semiwild pollinators. Between 120,000 and 200,000 species of animals perform this service.[18]

Bees are enormously valuable to the functioning of virtually all terrestrial ecosystems and such worldwide industries as agriculture. Pollination by European honeybees, for example, is 60 to 100 times more valuable economically than is the honey they produce. In fact, the value of wild blueberry bees is so great that farmers who raise blueberries refer to them as "flying $50 bills."[18]

While more than half of the honeybee colonies in the U.S. have been lost within the last 50 years, 25 percent have been lost within the last 5 years. Widespread threats to bees and other pollinators are fragmentation and outright destruction of habitat (hollow trees for colonies in the case of wild honeybees), intense exposure to pesticides, a generalized loss of nectar plants to herbicides, as well as the gradual deterioration of "nectar corridors" that provide sources of food to migrating pollinators.

In Germany, for instance, the people are so efficient at weeding their gardens that the nation's free-flying population of honeybees is rapidly declining, according to Werner Muehlen of the Westphalia-Lippe Agricultural Office. Bee populations have shrunk by 23 percent across Germany over the past decade, and wild honeybees are all but extinct in Central Europe. To save the bees, says Muehlen, "gardeners and farmers should leave at least a strip of weeds and wildflowers along the perimeter of their fields and properties to give bees a fighting chance in our increasingly pruned and ... [sterile] world."[35]

Besides an increasing lack of food, one fifth of all the losses of honeybees in the U.S. is due to exposure to pesticides. Wild pollinators are even more vulnerable to pesticides than honeybees because, unlike hives of domestic honeybees that can be picked up and moved prior to the application of a chemical spray, colonies of wild pollinators cannot be purposefully relocated. Since at least 80 percent of the world's major crops are serviced by wild pollinators and only 15 percent by domesticated honeybees, the latter cannot be expected to fill the gap by themselves as wild pollinators are lost.

Ironically, economic valuation of products as measured by the Gross Domestic Product fosters many of the practices employed in modern intensive agriculture and modern intensive forestry that actually curtail the productivity of crops by reducing pollination. An example is the high level of pesticides used on cotton crops to kill bees and other insects, which reduces the annual yield in the U.S. by an estimated 20 percent or $400 million.[18] In addition, herbicides used for a variety of reasons often kill the plants' pollinators needed to sustain themselves when not pollinating crops. Finally, the practice of squeezing every last penny out of a piece of ground by plowing the edges of fields to maximize the planting area can reduce yields by disturbing and/or removing nesting and rearing habitats for pollinators. With the above in mind, it seems obvious that the notions embodied in Eco-efficiency are hardly going to be effective in reversing the economic rationale supporting the processes that drive environmental degradation.

Unfortunately, too many people are fueled by their unquestioning acceptance of current economic theory, which not only designs and condones but also actively encourages the above-mentioned destructive practices. Such people simply assume that the greatest value one can derive from an ecosystem, such as a forest, is that of maximizing its productive capacity for a single commodity to the exclusion of all else. Single commodity production, however, is usually the least profitable and least sustainable way to use a forest because single commodity production simply cannot compete with the enormous value of non-timber services, such as the production of oxygen, capture and storage of water, holding soils in place, and maintaining habitat for organisms that are beneficial to the economic interests of people. These are all foregone when the drive is to maximize a chosen commodity in the name of a desired short-term monetary profit. Ironically, the undervalued, discounted, and/or ignored uses of the forest are not only more valuable than wood fiber production in the short term but also are more sustainable in the long term and benefit a far greater number of people.

For example, one study of alternative strategies for managing the mangrove forests of Bintuni Bay in Indonesia, a study more in keeping with the posits of the Genuine Progress Indicator, found that leaving the forests intact would be more productive than cutting them, according to Janet N. Abramovitz.[18] When the nontimber uses of the mangrove forests, such as fisheries, locally used products, and control of soil erosion, were included in the calculation, the most economically profitable strategy was to retain the forests. Maintaining healthy mangrove forests yielded $4,800 per hectare (2.5 acres) annually over time, whereas cutting the forests would yield a one-time value of $3,600 per 2.5 acres. Maintaining the forests would ensure continued local uses of the area worth $10 million per year and provide 70 percent of the local income, while protecting a fishery worth $25 million per year.

Another way landowners can make money from their forests without focusing solely on the cutting of timber is to use their forests for sequestering carbon.[36] In New South Wales, Australia, for example, David Brand, executive general manager of the state forests, watched the demand for timber declining, and in that decline he saw an opportunity to sell environmental services to local and foreign power companies that were looking for ways to offset the carbon dioxide their generating plants were releasing into the air. What would he sell? He would sell the sequestration of carbon (called "carbon storage rights") in the trunks and root systems of the forests' trees, for which he soon had an agreement to plant 2,500 acres of degraded pastureland in eucalyptus trees for $10 for each ton of carbon sequestered. In Japan early in 1999, Tokyo Electric Power Company signed a letter of intent to plant up to 100,000 acres of trees over the next decade. "We don't need to cut timber at all any more," Brand said. "Our forests are being driven completely by environmental values."

Selling carbon storage rights is a smart move because forests are increasingly recognized as a major factor in the reduction of carbon dioxide, the primary greenhouse gas of global warming. Creating a market for this service requires three main ingredients: a market framework, a demand from willing buyers, and a supply from willing sellers.

The three ingredients necessary to succeed at a significant scale are (1) formulating a framework of policy and political support to establish a level playing field that defines the commodities to be traded (and their varying qualities) and implements a system of credits and crediting that reduces financial risk; (2) creating a foundation of interested customers, which means educating people who want to reduce carbon emissions about the options that conservation of forests and sustainable forest management can fulfill as avenues for effectively reducing the amount of atmospheric carbon dioxide and thereby mitigating emission from generating plants; and (3) building the supply, which means helping private forest landowners understand the dynamics of carbon in their forests, how to increase the storage of atmospheric carbon in their forests, and how they can enter the carbon market with high-quality domestic carbon credits to sell.

To this end, the Pacific Forest Trust, headquartered in Boonville, CA, has analyzed carbon storage under four types of variable retention silviculture and compared them with clear-cutting, in which no carbon is stored. Results of the analysis show that an additional 32 million tons of carbon would be stored on a given site for over 50 years under variable retention harvesting. The analysis was based on three structural principles to ensure the credibility of the resulting carbon credits: permanence, additionality, and verifiability.

The foundation of the analysis is the *permanence* with which the carbon will be stored, which means one must assume that the gains in stored carbon are permanent by using such tools as conservation easements that would not only protect a forest from being converted to a nonforest use but also ensure that its management would be permanently altered to increase the storage of carbon. The latter would ensure that changes in future ownership would not reduce the gains in carbon storage. In Costa Rica, for example, high-quality carbon credits are currently derived from permanently dedicated parks and permanently secured conservation easements.

Additionality is a newly coined term that means a forest landowner must do something significant in addition to that which he or she is currently doing to ensure the trees on his or her property are increasing their storage of carbon, such as letting them grow for a notably longer period of time than was previously allowed before harvest. Additionality is also ensured by the conservation easement, which makes changes in management goals permanent and above prevailing norms. *Verifiability* is ensured as much as possible by using well-documented data on the forest type, state-of-the-art modeling based on decades of published scientific research, and an annual third-party assessment that is required by the conservation easement.*

Although selling carbon credits has the potential to help reduce atmospheric carbon dioxide as a greenhouse gas, we can no longer assume that the services Nature offers free for the taking are always going to be there because the consequences of our frequently unconscious actions affect Nature in many unforeseen and unpredictable ways. What we can be sure of, however, is that the loss of individual species and their habitats through the degradation and simplification of ecosystems can and will impair the ability of Nature to provide the services we need to survive with any semblance of human dignity and well-being. Losses are just that — irreversible and irreplaceable. To keep such things of value as Nature's inherent services, we must not only shift our thinking to a paradigm of sustainability but also calculate the full costs of what we do — Genuine Progress Indicator.

If the reductionist mechanical world view, as refitted with the aforementioned notion of Eco-efficiency, were replaced with sustainability, it would look something like this:

- *Eliminates* clear-cutting, except where ecologically necessary to create or maintain biological sustainability
- Measures prosperity by the *choices* saved and passed forward to the next generation and the *richness* of things from which to choose (natural capital) that accompanies those choices
- Measures productivity by the *ecological integrity and health* of one's forestland
- Measures progress by the *consciousness* with which one cares for one's forestland as a biological living trust as measured by the Genuine Progress Indicator
- *Integrates* aquatic habitats and riparian zones in the forestland as part of a seamless, interactive whole
- *Eliminates the notion of waste* by seeing everything in the forest as part of the recyclable, reinvestable biological capital that maintains forest integrity and productivity
- Sees the need for regulation as *failure* in forestland trusteeship
- *Honors and protects* biological, genetic, and functional diversity as the principal of the biological living trust in order to protect the productive capacity of a given forestland to provide a sustainable

---

* If you want to know more about the work of the Pacific Forest Trust on carbon storage, contact Laurie A. Wayburn at P.O. Box 879, Boonville, CA 95415 or call (707) 895-2090.

level of interest in terms of economic goods and services for the present and future beneficiaries

To achieve the kind of revolution in consciousness that is called for by the paradigm of sustainability, we would do well to heed an ancient Arab proverb as a point of departure: each word we utter should have to pass through three gates before we say it. At the first gate, the keeper asks, "Is this true?" At the second gate, the keeper asks, "Is it necessary?" At the third gate, the keeper asks, "Is it kind?"

How might this fit into caring for one's forest as a biological living trust? Each thought and action in caring for one's forest must pass through three gates: at the first gate of forest sustainability, the trustee asks, *"Is this which I am about to do ecologically sound?"* At the second gate of forest sustainability, the trustee asks, *"Is this which I am about to do necessary to the ecological integrity of the forest over time?"* At the third gate of forest sustainability, the trustee asks, *"Is this which I am about to do ecologically kind to the forest?"*

We have thus far spoken about certification in general terms. It is now time to examine certification in depth.

Chapter 3

The History of Forest Certification

The idea of forest certification (or an incentive-based tool for conservation) began with several organizations in the conservation movement and the wood industry at nearly the same moment in time during the 1980s. Although the initial attempts to use incentives, such as boycotts, to change forest practices were aimed at destructive practices in tropical forests, they did not have the desired effect. Consequently, the World Wildlife Fund, the Rainforest Action Network, and the Woodworkers Alliance for Rainforest Protection looked at promoting forestry operations that were both environmentally friendly and socially beneficial.

The Rainforest Action Network came out with a "Good Wood" guide, written by Pam Wellner and Eugene Dickey, that listed both "good" and "bad" companies, whereas the Woodworkers Alliance for Rainforest Protection sought to promote certain producers of tropical woods to their woodworking membership. The reasoning behind the action taken by the Woodworkers Alliance for Rainforest Protection was to reduce the harvest pressure on such potentially threatened species of trees as mahogany and rosewood. Along with these beginnings came the SmartWood program.

SmartWood

The SmartWood program was started in 1989 by Ivan Ussach, who was the director of the Rainforest Alliance, with the strong involvement of Dan

Katz, who was executive director.[37] They were joined by a number of outside advisors, such as Jack Putz from the University of Florida and Frank Sheridan of Afrasian Gross Veneers Ltd. The concept evolved as a consequence of a major conference organized in New York in 1988 by the Rainforest Alliance. The purpose of the conference was to examine the timber trade, boycotts of tropical timber, and forest conservation and management.

Based on the experience of the certification movement in organically grown foods, the Rainforest Alliance decided to test the concept of certifying sustainable forest management. After approximately a year of discussions and program design, the very first forest certification took place when SmartWood certified one of the largest reforestation programs in the world, Perum Perhutani on the island of Java in Indonesia.

The concept of the SmartWood Network was developed by the staff of the Rainforest Alliance in 1994 as a response to conditions the Alliance was finding through its certification work in Asia, the Americas, and Europe. The SmartWood Network came about because:

1. Region-specific, nonprofit organizations were requesting the Rainforest Alliance's technical assistance in developing programs with an explicit interest in long-term collaboration.

2. The Rainforest Alliance felt the need to forge a common bond between nonprofit organizations to ensure that certification was truly accessible to all types of companies, forest landowners, and organizations with strong commitments to environmental and social-community issues.

3. Despite hopes to the contrary, programs of region-specific nonprofit certification were finding it difficult to compete in the marketplace with large international certification programs (even the Rainforest Alliance was being put in the position of having to compete with regional nonprofit certifiers, a position SmartWood found unacceptable).

4. The demands of large international wood-processing companies needed the capability to assess the sources of woods in many regions.

5. There was a need for a cost-effective worldwide network to deal with the chain-of-custody once the timber had left the forest.

6. The desire of regional nonprofit organizations was to explore certification as a tool without committing their whole future to it because no one could be sure where certification was headed, which allowed organizations to move in and out of the network while at the same time ensuring that certified operations still had a system within which to maintain their certification.

With these considerations in mind, the SmartWood Network adopted a philosophy and set of operating principles.

The Philosophy and Operating Principles of SmartWood

In order for the SmartWood Network to be credible, the membership had to adopt a philosophy and operating principles that were ecologically sound, on the one hand, and other-centered (as opposed to self-centered) with respect to how clients were treated, on the other hand. We say this because people do not care how much you know until they first know how much you care about them.

Members of the SmartWood Network:

- Are committed to the necessity of a nonprofit, region- specific program that combines elements of research, education, and "on the ground" practical implementation of the best available scientific, social, and economic data
- Are therefore nonprofit, region-specific organizations
- Test certification as a tool for improving management on all forest lands, including those of indigenous peoples, publicly owned, and private nonindustrial
- Believe that certification must be for regional and national markets as well as for international and export markets, which implies a strong commitment to region-specific efforts of education and development of public policy that creates favorable conditions for credible forest certification and sustainable management
- Are committed to a policy of open access, which means that no applicant for certification will be turned away solely on financial grounds
- Will ensure that regional standards of certification are developed through an open and publicly transparent process of participation
- Will give strong support to certification of independent Council and its principles and criteria for sustainable forest management

A crucial part of the SmartWood philosophy is that solutions to problems surrounding sustainable forest management may be found in any individual, organization, sector, or culture regardless of religion, politics, training, or philosophical orientation. This philosophy places constant pressure on SmartWood staff and consultants to maintain an open analytical attitude toward the perspectives and opinions of others, which brings us to the Institute for Sustainable Forestry.

Institute for Sustainable Forestry

Before discussing the Institute for Sustainable Forestry, I (Walter), an ex-logger, would like to tell you why and how I became involved with the Institute for Sustainable Forestry and forest certification. My personal history is steeped in the timber industry because both my father and mother were employed in the timber industry for over 30 years. My father was a timber faller and later falling boss (bullbuck) for a locally owned timber company in northern California's redwood region. My mother worked intermittently in plywood and finger jointing plants.

I began working with my father as a bucker and apprentice timber faller in the early 1970s after a stint in college. It was during the end of the old-growth era in Mendocino County. For the most part, we were cutting the overstory of old trees left by the loggers during the first 100 years of logging. The old-growth trees were either "outlaws," trees that were not harvestable the first time around because they leaned heavily over a creek or draw and would break up when felled, were too dangerous to fall with old technology, or were of inferior grade (high-grading was the standard practice of "selectively logging"). Times were relatively good. I made a decent income for a laborer, saved money, and bought a small farm. During this time, California enacted forest practice regulations despite the fact that many of us in the timber industry believed that we were taking care of the environment and harvests were at a sustainable level — harvesting the "decadent" trees and promoting growth on the young trees.

The 1980s, however, brought a recession to the timber industry and the "good times" we had known were essentially over in more ways than one. The 1980s was the decade of unbridled greed within and without the timber industry and thus brought about the decay of the timber industry. Many family-owned mills went bankrupt. Large timber companies expanded their holdings. Logger and millworker wages and benefits were either cut or remained stagnant. Cutting timber outstripped the growth of trees on industry lands in my county by 300 percent, and harvest moved rapidly from old-growth trees to advanced second-growth trees that had regrown forests harvested 80 to 100 years earlier, to younger and younger stands of trees.

In short, it was the end of the boom times for timber families as sawmills shut down and employment plummeted. Between 1985 and 1992, 6 of the 11 sawmills shut down, including 4 owned by the largest forestland holder in the county. In addition, work shifts were reduced at mills owned by the second largest holder of forestlands.

In 1985, a partner and I started a logging company. Our desire (which was eventually accomplished by my partner) was to work for landowners who wanted to practice "good" forestry, which also meant "good" logging. The only steady work that we could get at the time, however, was with Louisiana Pacific, a large timber corporation, although we did manage to work for a few conscientious forestland owners. Our company included about ten seasonal employees besides ourselves and typical ground-based logging and road maintenance equipment: tractor, skidder, loader, grader, and water truck.

Our work with the timber company (Louisiana Pacific), although frustrating, was bearable for a while. The standard silvicultural prescription was shelterwood removal, where the majority of the trees were cut and removed, but some evenly distributed live trees were left standing. Unfortunately, that also included advanced regrowth trees (the ongoing joke at the time was we were to cut any tree that cast a shadow). All of us involved with this kind of "forest management" were frustrated, including many of the foresters who were told to make these kinds of cutting prescriptions. Forestry was not being practiced — log procurement for hungry sawmills and greedy owners was. In 1989, several changes occurred that ended my employment as a timber worker and began a transition to employment in sustainable forestry and forest certification.

Early in 1989, the CEO (Harry Merlo) of the timber company (Louisiana Pacific) responded in an interview with the regional newspaper; he said, "We don't log to a 6 inch top; we don't log to a 4 inch top; we log to infinity. We want it all now." As a result, my logging company began logging every stick of wood in the forest. We cut all tanoaks over 8 inches in diameter and skidded and loaded them full length; all conifers over 12 inches were cut and the tops were skidded and loaded as well. Any cull log or downed tree that was solid enough to skid was taken to the chip mill. This had gone beyond timber harvesting; this had even gone beyond timber mining. We were removing everything. "...it's out there; it's ours, and we want it now!"

In conjunction with the "scorched earth policy," the timber company was closing the local sawmills. Their rationale was not that they were running out of timber; it was those "damn environmentalists" stopping them from logging. By now, most of us realized that this complaint was a pure lie, but many felt powerless to do anything about it. Then the timber company opened up a resaw and drying plant in Mexico using machinery from a Mendocino sawmill. The company not only was ruining the forest, shutting down sawmills, putting people out of work, and blaming all of that on someone else but also was now exporting jobs to exploit the people of Mexico.

That was all I could stand. I went to a news conference with a bunch of environmentalists and labor leaders who were condemning the company's policies and actions. Of course, being a logger for the company, I made headline news — photos and all. Boy, was the timber industry mad!

Later that week I had a meeting with the company's area resource manager and chief forester. They basically said to put up or shut up, and I said there "ain't no way I'd shut up." As an aside, the two fellows who talked to me on behalf of the timber company were later fired, partly for telling the company that it was running out of trees.

I therefore sold my share of the logging company to my partner and began helping the local environmentalists make their case against the timber companies, both in court and in the regulatory arena. I also looked for work as a timber faller and discovered that it was hard to find a job. Finally, a small gypo-logger gave me a job and risked his reputation and possibly his livelihood because he believed that I had a right to say what I believed in, despite the fact that he disagreed with my position.

In 1990, I heard a man on the radio named Jan Iris talk about restoration forestry and local, value-added milling. I made arrangements to visit his operation. Here was a man and his wife, Peggy, who had a vision — a vision that opened a world of possibilities in my imagination. Included in their vision of forestry and local value-added manufacturing was the notion of a forestry certification program that was loosely based on organic food certification. I immediately knew this was the kind of positive, proactive program that I could fully be a part of. In addition, I had always felt that the environmental movement needed not only to work to stop forest destruction but also to provide an alternative mechanism for forest-based communities so they could continue their way of life and provide for their families.

About this time, I was also approached by a man named Hans Burkhart who had inspired the county of Mendocino to form a forest advisory committee that would analyze the county's forest resources and recommend policies that would provide for the long-term management of the county's forests. I was named by the county supervisors to this committee. Soon after the beginning of the committee's work, Hans bought me a book called *The Redesigned Forest* by Chris Maser. Consequently, my life made an incredible course correction, one born of new knowledge, and I hope there will be many others.

Ultimately, Jan, Peggy, and I, along with many others, founded the Institute for Sustainable Forestry (ISF) based on the Ten Elements of Sustainability, the foundation of the Institute for Sustainable Forestry's certification standards. I became the certification program director, and I wanted to take the idea of forest certification as far as it could go. Now it is time to look at the beginning of the Institute for Sustainable Forestry.

Jan and Peggy Iris, environmentalists and community activists in southern Humboldt County, California, ran a small sawmill and dry kiln that manufactured tanoak lumber. Tanoak is considered a "weed" by the timber industry to be exterminated because it possesses little or no commercial value and interferes with the growth of trees that do. Ironically, the same timber industry that would eradicate tanoak is responsible for the tremendous growth and spread of the tanoak woodlands because the industry cut and removed all of the commercially valuable conifers, such as Douglas fir and redwood, leaving tanoak free to grow and spread.

Jan and Peggy got their raw materials by practicing what they called "restoration forestry" on private lands in their area. The main objective of restoration forestry was to restore a battered forest to a more natural mix of species over time by using what they considered to be "light touch" selective silviculture. They also felt that good forestry was something around which their community could rally in the face of terrible conflicts within the community over the widespread practice of insensitive forestry employed by the timber industry.

They conceived an idea that loosely followed the certification programs for organic food, which would differentiate their forest products based on "restoration forestry" from the products extracted by the timber industry based on industrial forestry. They felt that a segment of the consuming public would support good forestry practices by purchasing such products, just as there is a segment of the public that supports good farming practices by purchasing certified organic foods.

In 1990, Jan and Peggy, with the help of the Southern Humboldt Economic Development Corporation, began holding meetings throughout the region with community members and owners of private forestlands. A meeting would be held once a month that would address some aspect of forestry or community stability, such as silvicultural practices, logging practices, stream protection, worker safety, or other issues surrounding employment.

A final meeting was held in January 1991 at which the basic elements of sustainability were developed. Over 50 community members, including forest activists, foresters, loggers, woodworkers, business people, and academicians, attended. During this meeting, the ten elements of sustainability were crafted, along with the certification program of the Institute for Sustainable Forestry.

The ten elements of sustainability are as follows:

1. Forest practitioners will maintain and/or restore the aesthetics, vitality, structure, and functioning of the natural processes, including fire, of the forest ecosystem and its components on all scales of landscapes and time.
2. Forest practitioners will maintain and/or restore surface and groundwater quality and quantity, including aquatic and riparian habitat.

3. Forest practitioners will maintain and/or restore natural processes of soil fertility, productivity, and stability.
4. Forest practitioners will maintain and/or restore a natural balance and diversity of native species of the area, including flora, fauna, fungi, and microbes for the purposes of long-term health of ecosystems.
5. Forest practitioners will encourage a natural regeneration of native species to protect valuable native gene pools.
6. Forest practitioners will not use artificial chemical fertilizers or synthetic chemical pesticides.
7. Forest practitioners will address the need for local employment and community stability; will respect workers' rights, including occupational safety, fair compensation, and the right of workers to collectively bargain; and will promote worker-owner operations.
8. Sites of archaeological, cultural, or historical significance will be protected and will receive special consideration.
9. Forest practices executed under a certified Forest Management Plan will be of the appropriate size, scale, time frame, and technology for the parcel and will adopt the appropriate monitoring program not only to promote beneficial cumulative effects on the forest but also to avoid negative cumulative effects.
10. Ancient forests will be subject to a moratorium on commercial logging, during which time the Institute for Sustainable Forestry will participate in research on the ramifications of management in these areas.

Over the next 2 years (1991 and 1992), the Institute for Sustainable Forestry developed its on-the-ground certification guidelines, which we will discuss in detail in Chapter 4.

After establishing its capacity for technical certification the Institute for Sustainable Forestry realized that, as a small locally based certifier, it had the expertise to evaluate regional forest practices and their social implications, but would have a difficult time making a significant impact in the marketplace. The Institute for Sustainable Forestry believed that a cooperative effort would attract greater public recognition, provide increased capabilities in marketing, enhance opportunities for public education, and make the Institute for Sustainable Forestry more competitive with other certification ventures.

For these reasons, the Institute for Sustainable Forestry joined the Rainforest Alliance's "Canada–United States Association of the SmartWood Network" and became the first member in 1995 by signing a cooperative agreement in June of that year. The Institute for Sustainable Forestry/SmartWood collaboration provided a model for the development of the operations of the Canada–United States Association of the SmartWood Network by certifying landowners in California under the Network's certification policies and procedures.

One of the SmartWood program's fundamental strengths and differences as a certifier is the collaboration with regional non-profit organizations to implement certification. The regional connections and knowledge these organizations bring to the SmartWood certification process are critical to the social acceptance and technical viability of the program. Members of the Canada United States Association of the SmartWood Network share a commitment not only to the responsible use of the forest but also to the protection, maintenance, and restoration of the forest ecosystem, while simultaneously sustaining rural communities and their economies. It is important to note that every organization in the Canada United States Association of the SmartWood Network operates other programs, besides certification, that address sustainable forestry and sustainable communities.

The Institute for Sustainable Forestry was the first forest management certification program in the world, and is still one of the few that was "home grown" and focused on the forests of home. How certification affects forest-based communities in California is the measure by which the Institute for Sustainable Forestry judges success. It is with this background that we will discuss the Institute for Sustainable Forestry's certification program and the larger certification movement, which brings us to the Forest Stewardship Council.

Forest Stewardship Council

The initial meeting of the Forest Stewardship Council took place in Washington, D.C. in 1992. At this meeting (which was called and paid for by the World Wildlife Fund and organized by Richard Donovan, currently the director of the SmartWood program), the Forest Stewardship Council was named, and an interim governing board was appointed. The two primary committees were the Principles and Criteria Working Group and the Organizational Development Group. These committees worked on their respective charges between March 1992 and November 1993.

The Forest Stewardship Council was officially endorsed in November 1993 by its membership at a founding meeting in Toronto, Ontario, Canada. At that meeting, two chambers (eventually broken into three chambers) were created to ensure the equality of representation among environmental, social, and economic interests. The Board of Directors was also subject to equal representation between the northern industrialized countries and the southern nonindustrialized countries. Criteria for membership were established and the membership endorsed the "Principles and Criteria" at a later meeting in 1994. An executive director was also hired in 1994 and an office was established in Oaxaca, Mexico.

To keep the Forest Stewardship Council balanced among the often competing interests — economic, environmental, and social — and their respective points of view, a three-chambered system of democratic governance was established. The following discussion of this balance is based on personal communication between Chris and F. David T. Arens, Secretary of the U.S. Working Group, Inc.

The chamber system of governance, which mediates the representation of members, is a fundamental organizational, structural, and philosophical underpinning of the Forest Stewardship Council. Through the chamber system, the Forest Stewardship Council carefully maintains an equal balance among its three constituencies — economic, environmental, and social — in its regional, national, and international levels of decision-making.

For example, in the Forest Stewardship Council-U.S. (which has bylaws modeled after those of the Forest Stewardship Council-International), the chamber system functions at several different levels. When a member applies to join the Forest Stewardship Council, the individual or organization is assigned to a chamber, which is based on the information provided in his or its application. Once a member belongs to a particular chamber, his or her vote on any matter for which a membership vote is required will be weighted to ensure that no chamber has more influence over a decision than another. Actions in which a membership vote is required include such things as election of Board directors, changes in the bylaws, or other matters put before a National Assembly for formal consideration.

Varying levels of chamber balance are also a mandated organizing principle for the Board of Directors in that each chamber must have three members on the Board. The Board's Executive Committee, if formed, must have at least one member from each chamber, as must the policy and advisory committees of the Board. Chamber balance is also customarily applicable to regional working groups, which are in charge of drafting forestry standards.

Below is a *hypothetical* illustration of how chamber balance works in voting.

> Each chamber has 100 total votes it can cast on a particular issue. Now, let's suppose the environmental chamber has 50 members, the economic chamber has 100 members, and the social chamber has 10 members. Let's suppose further that 15 environmental chamber members vote in favor of a particular issue and 35 are opposed; no social chamber members vote in

favor, 10 vote against; 70 economic chamber members vote in favor of the issue, 30 vote against.

Weighting of the votes:
Environmental: 100 votes/50 members = 2 votes per member
Social: 100 votes/10 members = 10 votes per member
Economic: 100 votes/100 members = 1 vote per member
Decision, based on weighting (must be a two thirds majority to pass on the first ballot):

Chamber	In Favor	Against
Environmental	15 yeas × 2 votes = 30	35 nays × 2 votes = 70
Social	0 yeas × 10 votes = 0	10 nays × 10 votes = 100
Economic	70 yeas × 1 vote = 70	30 nays × 1 vote = 30
Totals	100 weighted votes "yea"	200 weighted votes "nay" MATTER IS NOT PASSED Two thirds majority

Today, the Forest Stewardship Council accredits organizations to certify sustainable forestry provided they subscribe to the Forest Stewardship Council's "Principles and Criteria" and their procedural guidelines (to be discussed later in detail). Accredited organizations can be both nonprofit (such as SmartWood) and for-profit (such as Scientific Certification Systems); neither is discriminated against.

Things To Do Prior to Seeking Certification

A forestland owner should address the following issues before seeking certification:[28]

- Check the backgrounds of the respective certifiers; ask for references and follow through with reference checks.
- Be clear on the standards and criteria against which you will be evaluated and have the certifier specify them in writing; determine whether they are the official standards of the Forest Stewardship Council for your region.
- Require a clearly written description of the scoring process to be used in the assessment, including weights assigned to the standards (if any).

- Review the steps of the assessment that the certifier will follow; make sure everyone, including field staff, understands both the process and the time frame.
- Obtain a written cost estimate that covers not only the cost of the assessment itself but also the cost of annual audits and licensing under a certification contract.
- Ask what services the certifier will offer you if you proceed with its program, such as media coverage, market visibility, and assistance in networking.
- Ask who will be on the certifier's evaluation team and review their professional competencies.
- Require certifiers to clearly delineate the process for peer review of the assessment team's report; recommend peer reviewers.
- Consider a two-phased approach to the project: "scoping" to evaluate your readiness to engage in a full assessment, then a "full assessment," assuming the outcome of the scoping was favorable.
- Consider having official project observers; outside industrial and environmental representatives at the table can not only further the learning experience but also bring added credibility to the outcome of the assessment.

Having considered how the Institute for Sustainable Forestry fits into the SmartWood Network, how the SmartWood Network fits into the Forest Stewardship Council, and what questions to ask, it is now time to examine SmartWood forest certification as one certifier practices it, namely, the Institute for Sustainable Forestry.

Chapter 4

How the Certification Process Works

In this chapter, we will examine in some detail how the SmartWood certification process works as implemented by the Institute for Sustainable Forestry.[37] Making certification work as a tool of forest conservation and sustainable community development demands the cooperation and coordination of three major elements of the forest to the consumer chain of events. First, the forestland owner must meet the standards of certifications. Second, techniques for properly tracking products, such as timber, from the forest to the primary manufacturer, to the secondary manufacturer, to the wholesaler, and finally to the retailer must be in place to ensure that the certified product can be distinguished at every step of the way, from the noncertified product throughout the distribution system. Third, the consumer must be able to identify the product as certified and be willing to purchase it.

In the Forest

Applicant's Responsibilities Prior to the Arrival of the SmartWood Assessment Team

Before a full assessment team is assembled, the person who is applying for certification must fulfill the following nine conditons for the staff, either at SmartWood headquarters or the regional affiliate, such as the Institute for Sustainable Forestry:

1. The applicant must provide the following documentation:
   - Management plan, or for a particularly long planning document, copies of the plan's table of contents and the most important sections as determined through discussions with the appropriate SmartWood staff
   - Annual production in volume of timber, species of timber, and products over the past 5 years
   - List of primary personnel, those with seniority and those active in on-the-ground management
   - Organizational chart
   - List of people (name, phone and fax numbers) with some knowledge about the applicant; that is to include environmental organizations, government forestry agencies, community leaders, and academic institutions
   - List of sources from which products are obtained
   - Copies of the Forest Stewardship Council-endorsed or non-Forest Stewardship Council-endorsed guidelines for certification of which the applicant is aware in the region or in the country, if outside of the U.S. or Canada
   - If possible, a general map of the forest management areas
   - List of available documentation that, through a discussion with the lead person from SmartWood, is determined to be important enough to be duplicated and made available to the assessment team, either through the mail or upon arrival at the site

The information derived from the above list of items will give Smart-Wood the ability to initially assess the readiness of the applicant to move forward in seeking certification. Since certification is voluntary and has a financial cost attached to the assessment process, it is to the benefit of the applicant to know what potential impediments, if any, to certification might exist so they can be corrected before the full assessment proceeds. This gives the applicant time to learn more about certification and to decide if the timing is right to expend the necessary costs in time, energy, and money that certification requires.

2. Sign and return the contract agreeing to the assessment.
3. Make the minimum agreed upon advance payment to SmartWood.
4. Agree to the plan of work.
5. Notify relevant personnel, both internal and external, regarding the future arrival of the SmartWood team and proposed plan of work.

6. Assist in making appointments and scheduling as requested by SmartWood.
7. Arrange for hotel and, if necessary, for in-country transport and other logistics.
8. Make the necessary arrangements so that, on the team's arrival, the applicant can provide documents as follows:
   - All management plans
   - Comprehensive list of product buyers
   - Comprehensive list of the applicant's own properties and outside sources of products
   - Maps
   - Copies of written policies
   - Description of training programs
   - Copies of titles for lands under the applicant's management
9. May suggest persons to act as team members but has no vote in approving the people selected for the team; to this end, the applicant will be given a list of team members in order to provide his or her reactions and/or register their objections to specific individuals, but the applicant will not have the power to veto a selection.

It is preferred that all the documents listed above under no. 8 be duplicated and organized in folders or notebooks so each member of the SmartWood team has his or her own set for personal review. These documents are to be available upon arrival of the team so members of the team can get a preliminary "feeling" for the applicant's management operation.

On-Site Assessment Procedures

Members of the assessment team must meet:

- Internally among themselves to get to know one another and form the sense of a working team, which includes discussing the best strategy for designing a fair and thorough assessment in order to best serve the applicant and simultaneously uphold the credibility of the certification process
- With the applicant's operational staff to gain a sense of the internal operating philosophy and interpersonal relationships of the applicant's business
- With people outside of the applicant's business (such as personnel from local government, local environmental and/or community nongovernment organizations, and other people who "have their

finger on the pulse of the local situation) to gain a broad perspective of how people feel about the operation as a whole

Beyond this, the forest assessment process is generally divided into seven major operational phases:

Phase 1 is choosing the team.

Phase 2 is the initial team planning.

Phase 3 is the assessment of the applicant's management plan and management system.

Phase 4 is the actual assessment in the field.

Phase 5 is to assess information from other parties about the applicant.

Phase 6 is group analysis by the assessment team and its presentation of the preliminary findings, which is done on site before the team departs.

Phase 7 is writing the individual reports by the team members and the processing of the reports into a single report, which is peer reviewed by two or more nonteam members to judge its professional accuracy and apparent fairness after it is given in final form to the applicant.

Phase 8 is the decision of whether to certify the applicant's operation.

Phase 9 is the annual audit that follows the initial certification.

Phase 1, Choosing the Team

Assessment teams are chosen to assure that a balance of expertise covers ecological, social, and economic concerns. Depending on the situation, this multidisciplinary team usually has three or more members. One is likely to be a "dirt" forester who understands forestry production and economics. Others would include an ecologist, wildlife biologist, and sociologist. Team members are chosen from a list of people who have either taken the SmartWood Assessors Training Course or are known to SmartWood or a SmartWood Network member.

In addition, there is always a team leader who has not only gone through certification training but also has experienced being a team member on numerous certifications. The team leader is generally an employee of SmartWood or a staff member of a SmartWood Network affiliate.

Assessors are chosen not only for their expertise but also for their philosophical outlook on sustainable forest management. SmartWood generally stays away from extreme viewpoints. Either "all forestry is appropriate" or "logging should be banned" viewpoints can be problematic on an assessment team. Most importantly, potential assessors must be those who are willing to apply the standards to forestry operations without prejudice to who owns the operation or to the standards themselves. An assessor must be able to maintain confidentiality (within the context of the Forest Stewardship Council's public disclosure policies), be professional in manner, have good communication skills (both in writing and speaking), be analytical and adaptive, as well as be a good team player.

Having now served on a number of SmartWood assessment teams, I (Chris) am totally impressed with the fact that I have yet to experience a single clash of egos. Every team I have been on has been the epitome of the democratic principle in its highest form, which is one of the reasons I so love working with SmartWood.

Phase 2, The Initial Team Planning

Upon arriving on site for the assessment, the team must have one or more initial internal meetings. The purpose of the meetings is to do the following six things:

1. Discuss logistics.
2. Define members' roles and responsibilities, with a lead individual assigned for each subject in the guidelines.
3. Have the team leader present the certification process and strategy for the specific applicant and review the assessment schedule.
4. Review the guidelines and procedures for addressing each subject, its related criteria, and its scoring.
5. Review the list of contacts, meetings, and the interview process because it is not necessary for all individuals to go to all meetings, but it is necessary that all meetings are attended by at least one team member.
6. Adapt the SmartWood guidelines to meet the local conditions; for example, when in a country other than the U.S., it is good to have an in-country specialist act as a cultural liaison with respect to how SmartWood guidelines fit into the context of local customs and laws.

Phase 3, Assessment of the Applicant's Management

Phase 2 encompasses the initial meeting(s) with the applicant's staff and is designed to accomplish the following six tasks:

1. Review the schedule, confirm appointments, and get further suggestions of people to interview.

2. Have the applicant's staff verbally present their approach to forest management.
3. Discuss the methodology and logistics of the field assessment:
   - Plan assessment strategy with the applicant's managers.
   - In discussion with the applicant, SmartWood (not the applicant) selects which sites are to be visited.
   - Using maps, SmartWood will stratify the areas of forest management to select and visit an appropriate cross section of sites that are managed/harvested by different teams of people, sites in different forest types or different types of terrain, and so on.
4. Review the following documents prepared by the applicant:
   - All management plans
   - Comprehensive list of product buyers
   - Comprehensive list of the applicant's own properties and outside sources of products
   - Maps
   - Copies of written policies
   - Description of training programs
   - Copies of titles for lands under the applicant's management
5. Clarify any specific issues that have arisen during preparation for the assessment.
6. Ensure that field staff have the appropriate equipment and documentation during the assessment team's field visit.

Phase 4, Assessment in Field

All accredited certifiers, such as SmartWood, and their affiliated member certifiers, such as the Institute for Sustainable Forestry, the Rogue Institute for Ecology and Economy, and Scientific Certification Systems, must follow the same principles of the Forest Stewardship Council, in that they have no choice if they wish to remain accredited. In this section, therefore, we present the operating principles of the Forest Stewardship Council, which are as follows:

Principle 1: Forest management shall respect all applicable laws of the country in which they occur, and international treaties and agreements to which the country is a signatory, and comply with all Forest Stewardship Council Principles and Criteria.

Principle 2: Long-term tenure and use rights to the land and forest resources shall be clearly defined, documented, and legally established.

Principle 3: The legal and customary rights of indigenous peoples to own, use, and manage their lands, territories, and resources shall be recognized and respected.

Principle 4: Forest management operations shall maintain or enhance the long-term social and economic well-being of forest workers and local communities.

Principle 5: Forest management operations shall encourage the efficient use of the forest's multiple products and services to ensure economic viability and a wide range of environmental and social benefits.

Principle 6: Forest management shall conserve biological diversity and its associated values, water resources, soils, and unique and fragile ecosystems and landscapes, and, by so doing, maintain the ecological functions and the integrity of the forest.

Principle 7: A management plan, appropriate to the scale and intensity of the operations, shall be written, implemented, and kept up to date. The long-term goals of management, and the means of achieving them, shall be clearly stated.

Principle 8: Monitoring shall be conducted, appropriate to the scale and intensity of forest management, to assess the condition of the forest, yields of forest products, chain of custody, management activities, and their social and environmental impacts.

Principle 9: Management activities in high conservation value forests shall maintain or enhance the attributes that define such forests. Decisions regarding high conservation value forests shall always be considered in the context of a precautionary approach.

Principle 10: Plantations shall be planned and managed in accordance with Principles and Criteria 1 through 9, and Principle 10 and its attendant Criteria. While plantations can provide an array of social and economic benefits and can contribute to satisfying the world's needs for forest products, they should complement the management of, reduce pressures on, and promote the restoration and conservation of natural forests. (With the exception of the southeastern U.S., principle 10 applies primarily to countries outside of the U.S.)

In addition to listing the above principles, we will discuss how one affiliate member of the SmartWood certification network, namely, the Institute for Sustainable Forestry, applies local criteria in order to assess the extent to which the intent of the Forest Stewardship Council's Principles and their attendant Criteria are enacted on the ground when someone applies for certification of his or her forestland. We do this to give you, the reader, a concrete idea of how an actual assessment might work because such an assessment is the fundamental vehicle through which an affiliate member of the SmartWood certification network (or that of some other certifier) recommends either approval or denial of certification to their headquarters, where the decision is made whether to certify or not.

However, assessment of the forestry practices in the field takes place only after the initial review of the management plans, meetings with the applicant and his or her staff, and sometimes meetings with other affected parties (as described in phase 4). Prior to these meetings, however, the lead responsibilities for the three major facets of the assessment are divided among the team members: (1) sustainability of forest management and timber production, (2) environmental issues, and (3) socioeconomic issues. This is not to say that team members cannot comment on all facets of the assessment; in fact, some comment is usually made on areas other than the one for which a team member is primarily responsible.

The team members must have access in the field to the following documents:

- Assessment Methodology of the SmartWood guidelines, which is an expanded version of the generic SmartWood guidelines, including criteria, indicators, and assessment techniques for evaluating the applicant's forestry operation
- All pertinent maps
- Site practice record books, when available

Ideally, an assessment of forest management operation is like a financial audit in that the business being assessed is expected to present records for the auditor to spot check and verify the accuracy of the recorded details about the management operation. SmartWood therefore encourages the staff of management operations (especially the large and more sophisticated ones) to keep record books or data sheets in which to archive information about a harvest area, including information about pre- and post-harvest activities.

For example, the tree marker may keep records of each day's marking, which is signed by the marker. Such entries must include number of trees, species, and comments about trees marked for removal. In some cases, information may also be recorded about the recommended or designated direction in which a specific tree is to be felled. Then, during the actual harvest, accuracy of subsequent felling can be recorded against the recommended direction of felling. Similar records might also be kept for the results of the post-harvest inventory, assessment of damage to trees left standing, and activities surrounding closure of the site.

In addition to the above items, airplane over-flights are extremely important and strongly recommended for large commercial management operations, but not for small ones. Such over-flights offer one of the best ways of seeing an entire concession or management area and serve as a means to:

- Assess impacts of current and past logging.
- Confirm that unauthorized large clearings are not being created.
- Judge the quality of the forest resource in future management areas, as a basis for continued operations.
- See conservation zones and sensitive areas, such as swamps and steep slopes, that are excluded from management.
- Assess landscape level conservation issues.
- Assess whether there is encroachment into the management unit by other loggers, slash and burn farmers, or miners.
- Assess potentially relevant activities on neighboring properties, especially as they might affect ecological diversity on the landscape scale.

Assessment of aerial photos and satellite images is also recommended where these images are available or can be acquired with reasonable cost and ease by the applicant or the assessment team.

The assessment team is to enter the field only when it has a working knowledge of the forest management, methods of processing the forest products, and a clearly defined assessment strategy (Photo 9). The field assessment, based on the SmartWood guidelines, must address the adequacy of the applicant's forest management from nine different perspectives.

To begin with, the person or persons assessing the forestry practices portion of the business, which is the first portion of the guidelines, must have in hand a tape to measure tree diameter at breast height (dbh), 50-meter measuring tape, clinometer, prism, and a compass. Team members must conduct field work at sites that represent different stages in forestry activities, which can generally be defined as:

- Pre-harvest, which represents operational planning, such as sub-block layout, pre-harvest inventories, tree hunting and marking, skid trail layout and marking, and in tropical areas the cutting of climbing vines
- Harvest, which represents skid trails and hauling roads, felling trees, skidding trees, and the yarding of trees
- Post-harvest, which represents an assessment of the damage to leave trees, near-term silvicultural activities, such as the planting of trees, and the closure of roads and logging compartments
- Long-term activities (management, research, and tree regeneration), which represents such long-term silvicultural activities as thinning trees to release the chosen "crop trees" for faster growth over time, long-term monitoring for information on growth and yield, and research

Photo 9 Assessors in the field with forest managers looking at stream and riparian issues. Members of the team include an ecologist, sociologist, forester, and harvesting specialist — the Philippines. (Photograph by Anabel Garcia.)

Given the seasonality of forestry practices, it is not always possible to see all stages of forest management happening simultaneously, although seeing all stages of forest management as it is happening is the goal of the assessment. This goal can be approximated by visiting sites that in the collective reflect the different stages of management. Teams must therefore visit the following sites:

■ Current harvest, which must be the primary focus of all assessments because such sites provide the clearest indication of the quality of the management planning, operations, and environmental impacts
■ Recent harvest (1, 2, or 3 years old), which will enable the assessment of past practices and their effects on the environment, as well as the potential for regeneration of the forest stand; it will also enable the assessment of some post-harvest activities and long-term activities (such as silviculture, research, and regeneration) while providing an opportunity to evaluate whether there is multiple reentry into old management areas before the next defined cutting cycle, which demonstrates how well the applicant is following his or her own management plan

- Past harvest (5 and 10 years old), which enables the assessment of past practices and environmental effects, as well as the potential for regeneration; past harvest sites, like recent harvest sites mentioned above, will also provide an opportunity to evaluate whether there has been multiple reentry into old management areas before the next defined cutting cycle, which demonstrates how well the applicant is following his or her own management plan
- Research, such as growth and yield plots, should always be visited even if it occurs outside of the forest management area
- Sites of future harvest that are under pre-harvest planning in order to compare the forest prior to disturbance and to see evidence of the pre-harvest planning activities

Additional methods of stratified sampling are also recommended as feasible, such as the assessment of activities in different forest types, under different land ownerships, by different forestry teams in the field (fellers, skidders, supervisors), close to and far from established roads, and where harvesting has occurred in different seasons. It is also recommended that special attention be devoted to historically challenging or problem areas, such as steep slopes, difficult soils, and so on. Having discussed prefield activities, it is now time to turn our attention to the field assessment itself, which brings us back to the assessment criteria.

As stated above, the reason we have included the local adaptation of Forest Stewardship Council's Criteria by the Institute for Sustainable Forestry is to give you, the reader, a sense of what really happens on the ground. Without at least reading about the way in which one local affiliate applies criteria in the field, it is impossible to fully appreciate the thoroughness with which a field assessment is conducted. This said, however, it is imperative to understand that such criteria are constantly being improved as assessors and affiliate certifiers gain both experience and an increasingly better understanding of local and regional conditions. Therefore, when you read the following criteria, keep in mind that they already may have been modified to some extent in order to improve them and to keep them current with any improvements made by the Forest Stewardship Council to their overarching Criteria. In addition to the above, we have added a few examples from other parts of the world to illustrate that critera are adaptable.

1 — Compliance with Laws and Principles of the Forest Stewardship Council

1.0 — Commitment to Principles and Legal Requirements of the Forest Stewardship Council

Since SmartWood is accredited by the Forest Stewardship Council, all certified operations must logically demonstrate a commitment in policy and practice to the Forest Stewardship Council, central defining document, "Principles and Criteria for Forest Management." The guidelines of both SmartWood and the Institute for Sustainable Forestry have been designed to meet the intent of the principles and criteria of the Forest Stewardship Council and, in fact, follow the same format and structure. An operation certified under these guidelines would logically have to show its commitment to the principles and criteria of the Forest Stewardship Council. Managers of forestry operations with large landholding may certify a portion of the property to "test" the certification process. They would, however, be expected to illustrate their commitment to the principles and criteria of the Forest Stewardship Council by certifying additional portions of their landholdings over time.

Operations certified by SmartWood must also be in accord with national, state, tribal, and/or local laws. The purpose of the certification process is to check with government agencies and other concerned parties to verify that an operation is dealing with legal requirements in a responsible fashion, which can be a valuable, nonthreatening way of helping operations improve the quality of their compliance. Assessment of actual legal compliance, however, is the mandated task of governmental institutions. In some cases, there may be applicable international conventions or treaties that also apply, as is clearly the case for endangered species under the Convention on Trade in Endangered Species. SmartWood assessors are responsible for pointing out what they perceive to be conflicts between laws, the guidelines, and international treaties or conventions, although in practice this has rarely occurred.

One of the instances in which conflicts may arise between laws and the Forest Stewardship Council's "Principles and Criteria" is where the government owns and leases forestland concessions. Some governments mandate the amount of timber that can be cut annually, and some also mandate the silvicultural responsibilities to the concessionaire. If Smart-Wood finds that either is out of compliance with the standards, it would be necessary for the concessionaire to request the particular government to change its law if certification were to become a reality.

It is important not only that forestry operations follow local, national, and international laws and the best available management practices but also that they meet their intent as well. There are a few states in the U.S. and numerous other countries that have extensive laws that regulate what is and is not permissible in forestry practices. Through these laws, citizens and legislators try to promote good forest management practices by regulating behavior that is considered bad or unacceptable in order to foster behavior that is perceived as good or acceptable.

Unfortunately, those who would rather work outside acceptable community definitions of good forest management continue to do so under most laws. They can continue to operate at the edge of the law because they follow the letter, but not the intent, of the law, which is to say that regulating forestry operations is difficult at best because many of the laws are more or less ambiguous, rather than black and white.

To enforce laws regulating forestry operations, the enforcement agency must prove not only that the law was broken but also that its breaking was intentional. Certification, on the other hand, is mostly concerned that the intent of the law has been met, meaning that operations are attempting to manage in such a way that their management responds to the public's desire for sustainable forestry.

The criteria dealing with laws and treaties should not confuse the fact that certification is a voluntary process. Certification is a market-based approach that allows consumers to demonstrate their support for well-managed forests by purchasing wood products that carry an accredited label and/or the Forest Stewardship Council's label. Certification is therefore an incentive approach to improving forest management rather than a regulatory approach that in essence punishes bad behavior when the perpetrator is caught.

1.1 — The forestry operation meets local, state, and national laws concerning the environment, forestry, and labor.

1.2 — The operation is up-to-date in its payment of local taxes and/or timber rights.

1.3 — Managers are aware of applicable international conventions, agreements, or treaties (including relevant treaties with indigenous Americans) and provide guidance so that operations in the field meet the intent of such international conventions, agreements, or treaties.

1.4 — A manager(s) is willing to make available to the public a summary of the certification of his or her operation.

2 — Forest Security

2.0 — Land Tenure and the Rights and Responsibilities of Land Use

Consistent long-term forestry will occur only if a manager(s) can be relatively certain that the forest in question will remain as forest and that

**Photo 10 Forest protected by community-based forest management group —
the Philippines. (Photograph by Walter Smith.)**

the manager(s) and users have clear rights and responsibilities. This is
important because many parties and factors can influence the tenure of
land and the rights of users. The intent of this section is to ensure that a
candidate operation is taking realistic steps to protect and maintain the
forest over the long term, including resolving conflicts with its neighbors
and users of the forest (Photos 10 to 12). In some cases, this may mean
protecting the forest from threats of competing uses or misuses, such as
unplanned, unauthorized, or illegal hunting, trapping, or fishing; illegal
logging; unplanned, unauthorized, or illegal grazing; and so on (Photo
13). In other cases, the manager(s) may take precautionary steps to
improve forest security by negotiating with local communities and/or
individuals to help control joint access to the forestland and its resources.

Land tenure and the rights and responsibilities of land use are extremely
important for long-term social-environmental sustainability. The owner of
forestland has a great responsibility to future generations to assure that
the land is managed in a biologically sustainable manner. Long-term forest
management cannot take place unless forest managers make sure that
forestland will stay in a forest condition and that the rights and responsibilities of forest managers and other forest users are clearly spelled out.

**Photo 11   Forest protected by community-based forest management group —
the Philippines. (Photograph by Walter Smith.)**

The clarity of these rights and responsibilities means protecting the
forest from threats of competing land uses — or misuses. Misuses are
such things as overharvesting, timber trespass, illegal hunting, and so on.
Threatening uses may be extenuating circumstances that necessitate the
land be subdivided because of financial obligations, such as inheritance
taxes, family health problems, or through loss of control by the sale of
the land to other parties. Forest managers must therefore take steps to
improve the security of their forestland by carefully negotiating joint
agreements for management or access to forest resources with individuals,
local communities, or companies and by providing the appropriate parties
with information about financial protection.

Be that as it may, much of the forestland in the world is owned by
governments, which lease it to companies or communities to manage it
for its resources. In many cases, particularly in non-industrialized countries, conflicts arise over the rights of use between the lessee of government forestland and the local citizens, particularly if the latter are not
given a fair opportunity to participate in and/or profit from the forest
management operations.

I (Walter) have evaluated operations in non-industrialized countries,
where I have experienced two clearly different approaches to forest

Photo 12   If the forest does not sustain the peasants of the Philippines, corn and beans from slash and burn agriculture will sustain them for a time. (Photograph by Walter Smith.)

management, both of which affect forest tenure, the rights of use, and issues of security, that in turn affect the long-term sustainability of the forest in question. In one instance, for example, a company managed land that had unresolved community land claims as well as needed local benefits from forest uses. Both the company and government refused, however, to recognize either the rights of use by members of the local community or their land claims, which together were critical to the community's well-being because there was neither a social nor an economic safety net to provide any kind of welfare. Management in this case was neither sustaining the social fabric of the local community nor protecting the forest environment.

The lack of community recognition was responsible for a substantial amount of timber theft by members of this selfsame community (Photos 14 and 15). The stolen wood was sold to traders on the "black market." It is tempting to condone such thefts when local families need the income but are denied all rights of use as well as jobs in the local forest.

Some such conflicts have even led to managers and community members killing one another over unresolved rights. In addition, the theft of timber can be great enough to damage the forest environmental itself.

Photo 13 A controlled access road; note the well rocked road and bridge crossing, the latter taking the place of a culvert — Mendocino County, California. (Photograph by Walter Smith.)

In the other case, a community-based forest management agreement was entered into between the government and local communities surrounding a 35,000-acre concession area. The production-sharing agreement with the government, in this case, is good for 25 years, with an optional renewal for another 25 years. The communities were given the rights and responsibilities to manage and use the forestland and its resources.

The communities thus formed a cooperative, which is operated by an elected Board of Directors with equal representation from the three main communities in the concession area. When in full operation, 220 workers are employed, including 25 regular workers, 10 management staffers (manager, forester, and finance/accounting staff), and 15 forest guards. The others work in the cutting areas, sawmill, and consumer store.

The Board of Directors approves both the number of people to be hired and the wage rates. The general manager makes the selection of workers based their qualifications and endorses their hiring to the Board. For laborers in harvesting, the selection is usually facilitated by the members of the board representing the community where the cutting area is located. There is a hiring preference for members living near the cutting areas.

Photo 14 Theft of timber devastated this hillside forest in the Honduras. (Photograph by Walter Smith.)

Indigenous peoples and settlers within the concession area are permitted by the cooperative to use nontimber forest products for commercial and traditional use. There appeared to be no barriers for women to overcome in running the cooperative; out of its 393 members, 116 are women. Two of the Board members are women, and 50 percent of the regular (office) workers are women. In addition, the cooperative programs for women include raising seedlings in the nursery and loans for handicrafts. The cooperative also plans to support a day-care center.

It appears that this highly participatory and democratic process has all but eliminated in-migration, settlement, slash and burn agriculture, illegal logging, and encroachment on heretofore unentered forest. In addition, nearly everyone's necessities are being addressed and met.

In the U. S., on the other hand, the biggest threat to land tenure and long-term sustainability is the transfer of land to a new ownership, inheritance taxes, poor financial planning, or some combination of the three. An unusual and innovative solution to these problems has been implemented by a family-owned company in the southern U.S. that both manages forestland and operates a sawmill.

The family is in its third generation of ownership, with the fourth generation employed at the company, learning the business and waiting in the wings to take over. Although the company itself owns only a modest

Photo 15 These four stolen logs are worth almost a year's income for a poor peasant family in Indonesia. (Photograph by Walter Smith.)

amount of forestlands, it manages 125,000 acres of nonindustrial forestland for over 50 families in the area, some of whom have been with the company from the beginning days of its forest management. How did something like this come about, you might ask.

The company came to the realization over 40 years ago that their sawmill business would survive from one generation to the next only if they could procure a sustainable supply of high-quality wood. They also understood what sustainable forests meant to the survival of their community and its surrounding environment.

The company foresters, which include father (owner) and son, manage nearly every aspect of the forestry and land-use requirements for the individual landowners — everything from boundary surveys to silviculture, harvesting, merchandising, monitoring, patrolling, supervising hunting clubs, and other activities. The forestland owner is required by contract to relinquish all forestry decisions to the company, which allows for consistency of management activities on the ground. The company, in turn, practices all-aged, all-species management, with cutting cycles that promote the establishment of large diameter, older trees. This meets both the objective of having a diverse forest with structural attributes not generally found in the area and producing high-quality sawlogs.

The company, for its part, has a formula based on the price of wood reported quarterly in a trade journal, by which it determines the price paid the landowner for his or her timber, which helps alleviate the appearance of a conflict of interest. Furthermore, the company does not charge the landowner for its forestry services.

When a landowner experiences a cash shortfall, whether from a family illness, college-bound children, inheritance taxes, business failures, or any other reason, the family looks to the forest as a way of capitalizing its shortage. The company, however, is so determined to manage the forest according to what the forest can provide that, if the landowner needs income before the next scheduled harvest, the company will pay the landowner for the timber before it is cut — sometimes several years in advance. The company also helps landowners in the same way with estate sales. Finally, if the land is being purchased by an outside party, members of the company work with both seller and buyer to keep the forest management intact by providing financing help, in addition to working with both parties to transfer the forest management services to the new owner. It is this kind of continuity that the following critera are designed to encourage.

2.1 — Land tenure is clear and legally secure (title, lease, marked boundaries, and so on).

2.2 — The owners have dedicated the land to long-term forestry (estate plan, easement, management plan, controlled rights of use, and so on).

2.3 — Conflicts with adjoining landowners or other users concerning resources are addressed and resolved in a systematic and legal manner.

2.4 — Unauthorized or illegal trespass, such as hunting and theft of timber, is controlled.

3 — The Information Base, Planning, and Strategy for Management

No. 3 assesses the consistency of the management strategy with the philosophy of the SmartWood program, as well as the adequacy of the available information and the management plans.

3.0 — Forest Management Planning

Planning forest management is a process, not just a document. A written plan is a concrete tool that improves understanding and coordination of the management approach by staff, contractors, and other interested parties. A written plan also facilitates consistency in the face of changes in personnel, landowner, and so on. Size and location of the forestland are extremely important in determining realistic biological, social, and economic expectations from implementing the management plan. The absence of a written forest management plan means that an operation cannot be certified, except for the following cases:

1. Where significant documentation already exists that meets most, if not all, of the data requirements of a management plan and virtually the only step remaining is to compile, formalize, and produce an overall document
2. Where the mere completion of a written management plan will have no major effect (positive or negative) on the quality of the field operations with respect to environmental, silvicultural, or socioeconomic practices
3. Where there is a well-documented system of forest management at a general level that provides clear guidance and consistency for site-specific management practices (generally relevant only for small properties managed by the same consulting forester or company)

Although the above situations do not eliminate the need for management planning, SmartWood emphasizes performance in the field over documents. While this does not reduce the value of documented management planning because experience clearly shows they are unequivocally valuable, the question is one of balance. In the SmartWood assessment process, on-the-ground performance is regarded as "the first among equals."

SmartWood expects management plans for large operations to be more detailed and systematic than those of small ones due to the financial constraints of small landowners and the relatively larger risk of negative environmental effects by large operations. As greater understanding of the importance of landscape level biological concerns has been realized, increased emphasis has been placed on this topic during assessments, particularly for medium and large forest holdings. Although concerns with and for adjoining landowners are always important, no matter what the size of an operation might be, expectations in terms of local consultation with neighbors are clearly higher for large operations, both during and after the initial planning process.

The Institute for Sustainable Forestry/SmartWood does not advocate any single silvicultural approach (e.g., even-aged vs. uneven-aged, single tree selection vs. group selection, and so on). Rather, certified managers are expected to balance production with environmental protection, weigh the advantages and disadvantages of the various forestry practices accordingly, and select those that maintain or restore ecosystem integrity while simultaneously responding to social and economic necessities. Because every practice can be used wisely or abused, experience (both certification and management) indicates the necessity of self-monitoring within a forestry organization to provide internal quality control, identify operational challenges, and report on the success or failure of management practices to solve problems.

We hope it is clear that one of the philosophical underpinnings of SmartWood and Forest Stewardship Council certification is that forests must be managed as a whole living system, not just for timber. A good management plan provides enough information and guidance to ensure the protection of all components of the ecosystem.

A written plan provides a number of essential elements to long-term forest management and must therefore be seen as a living process, not just a time-encapsulated document. As a living process, a forest management plan improves communication with management staff, landowners, and other observers. It documents the landowners' and/or land managers' vision, goals, and objectives, which can be articulated to and maintained by future family ownership, new personnel, new landowners, and so on. It becomes at once a document that provides a history of actions and results from planning decisions and helps to track changes in the forest as a result of management practices.

The size of the ownership and its location (i.e., country) are extremely important in determining expectations in terms of management planning. SmartWood and the Forest Stewardship Council expect that management plans for large operations will be more detailed and systematic than those for small landowners. Large landowners benefit from economies of scale, which allows them to make greater investments in data collection and analysis than is possible for small landowners. Large landholdings also require more intensive planning to coordinate activities and track the effects of management across larger landscapes.

In this respect, monitoring is essential for adaptive management. Continuous monitoring of resources allows managers to maintain up-to-date estimates of sustainable harvest levels, calculate present and future values of commercial species, assess cumulative environmental effects, and prioritize management activities. Monitoring is also an important tool for communication in that written results provide evidence of good management to concerned parties, such as local communities, environmentalists, and regulatory agencies.

A complete monitoring program involves both systematic data collection and on-the-ground observation. Smaller properties regularly visited by landowners and/or forest managers do not require complex, formal monitoring programs to track cumulative effects across the landscape. But the larger the forest property, the more detailed and systematic the monitoring protocol must be.

It has been my (Walter's) experience, however, that many of those landowners, companies, and/or land managers who possess the most incredibly sophisticated forest management planning tools (e.g., computer simulation software, Geographic Information Systems (GIS), satellite imaging, and so on) generally have the most difficult time meeting the standards of certification for forest management practices. On the other hand, many of those landowners, companies, and/or land managers with the least amount of formal written information and data have forest management practices that more consistently meet the standards of certification.

There may be several reasons for this. First, those who do the least amount of work on paper and/or computers are more consistently out on the ground actually doing the hands-on work, spending time observing, and making mental notes of changing environments. Second, as the saying goes, "garbage in, garbage out," which means that not enough time is spent "ground truthing" the computer-generated information. And third, some of the highly technical documents are produced as "window dressing" to impress potential buyers of the forestland, to prove to regulators that the company is in compliance with laws, and are used as advertising to impress the public, but are not used as real documents for planning and monitoring.

A number of the highest quality certified forest management companies in the U.S. had preconditions or conditions placed on their certification because their management plan was in their heads instead of on paper. These companies tend to have single ownerships that spanned several generations and have foresters who spent most of their time in the field, rather than behind a desk. In addition, I (Walter) have observed that foresters like working for such companies so much that many had been with the company a long time and had apprenticed with a previous forester that had also been there a long time. Information was therefore continually past from one generation of forester to the next without a break.

One of the more interesting and deliberate management plans was created by the cooperative described in 2.0 above. The cooperative developed a long-term written plan through a consultation process with members of the respective communities. The cooperative formed a planning committee composed of sub-committees, which includes subcommittees for research, inventory, processing, and livelihood. Others involved in the planning process are community leaders, members of the local municipal government, forest managers, and resource agency staff. The plan not only includes a description of the area and the conditions of its resources, socioeconomic issues, needs of the respective communities, opportunities for employment, common enterprise activities, land-use plans, and maps but also has a vision statement, goals, objectives, and strategies for implementation of forest management.

The management plan states the following goals: (1) to uplift the socioeconomic conditions of the communities and to provide them technical knowledge in different livelihood projects introduced; and (2) to rehabilitate the open and denuded areas within the concession area and to manage the forest sustainability. The stated objectives are to (1) maintain and protect the area classified as Protection Forest, (2) establish 1,002 hectares of plantation areas, (3) establish agro-forestry, (4) establish different types of livelihood projects, (5) utilize 67,500 cubic meters of timber, (6) utilize about 15,000 linear meters of minor forest products for a 5-year period, and (7) conduct boundary delineation of concession areas.

Planning forest management is thus a question of balance between performance, documentation, and the health of ecosystem processes. In a SmartWood assessment, on-the-ground performance might be regarded as "the first among equals," as will be seen thoughout the rest of this section on forest management planning.

3.1 — A multiyear forest management plan and/or other documents are written and available.

3.2 — The forest management plan is comprehensive, site-specific, and includes the following elements:
  i. A clear statement of vision, goals, and objectives
  ii. A description of the timber and nontimber resources being managed, environmental limitations to management, and land uses
  iii. A description of and rationale for the silvicultural system chosen and for the annual or periodic harvest level, including measurements to ensure adequate regeneration of commercial species

iv. A description of future forest condition that can be expected to result from the proposed management practices

v. A description of and rationale for the harvest system chosen and the equipment proposed for use

vi. A map that includes pertinent items, such as management units, forest types, harvest areas, conservation zones, research areas, roads and landings, and infrastructure

vii. A description of strategies to protect the forest from and restore it after such events as fire, infestations of harmful insects and disease, and human encroachment

viii. A description of measures to protect such environmental features as soil, water catchments, riparian areas, habitat, diversity (biological, genetic, and functional), and the environment against chemicals

ix. A description of financial projections, utilization of forest products, and marketing strategy

x. A description of the formal and informal process of consultation with neighbors, affected parties, and/or customary rights of use or lease agreements to forest resources within the property

xi. A description of monitoring protocol that shows how management practices will be adjusted based on new ecological, silvicultural, and/or socioeconomic information

3.3 — Nontimber forest products to be harvested have been inventoried and their management is appropriately incorporated into the planning process.

3.4 — Maps and work plans are produced at a useful scale for the supervision and on-site monitoring of management activities.

3.5 — Topographic maps are used to specify areas that are suitable for all-weather or dry-weather timber harvesting; location of roads, skid trails, and log decks; drainage structures; and buffer or conservation areas before logging and/or road construction begins.

3.6 — Clear guidance is given to field staff and contractors (in the form of written plans, manuals, and/or maps; clear policy directives and/or training) so they understand and implement the letter and the spirit of the forest management plan.

4 — Sustainability of Forest Management

The criteria in this section address the sustainability of the management strategy and the silvicultural prescriptions from the perspective of long-term forest production with a minimal biophysical effect on the forest ecosystem. This group of criteria addresses specifically the technical quality of how the forest management is implemented, including planning, harvesting, skidding, and post-harvest practices.

4.0 — Forest Management Practices

Forest management practices for certified operations must demonstrate, on a daily operational basis, consistent and balanced attention to environmental, silvicultural, and socioeconomic conditions and priorities. SmartWood therefore encourages forestry practices that produce a sustainable output of high quality wood while simultaneously protecting ecosystem process and the free biological services they provide. This means that management practices must provide habitat for a diversity of species, promote soil fertility, and protect water quality and hydraulic processes (Photo 16). Such practices must also maintain a complex forest structure that includes large standing dead trees or snags, hollow den trees, large downed wood on the forest floor, and well-managed riparian corridors in addition to quality trees for producing lumber.

Because SmartWood's philosophy is one of planning *now* for the present and the future, the harvest of timber is not allowed to exceed the growth of the trees. On ownerships with depleted stands of trees, this means that levels of harvest may be set below the potential growth of a stand in order to rebuild what is called "standing inventory." The rule of thumb in managing in the present for the present and the future is to remove trees across size and height classes, initially favoring thrifty, well-formed, fast-growing trees of all species to create ecological conditions that will foster long-term forest health and economic viability. Such management, in the Pacific Northwest, may include the creation of small openings in forest types that are typically managed in an even-aged plantation mode by planning the size and sequence of openings and the creation of forest edges while simultaneously accounting for landscape-scale ecological processes.

In many areas, past logging practices have left a legacy of roads in and adjacent to streams, draws with soil and organic debris, stream channels eroded down to bedrock, streams diverted by roads, wasting gullies, eroded soil, and degraded aquatic ecosystems. Forest trusteeship not only requires rehabilitation of such degraded areas but also management to avoid similar problems in the future. This said, the largest and

Photo 16   Selective logging with a small yarder. (Photograph by Jude Wait of the Institute for Sustainable Forestry, Willits, CA.)

most lasting effects of logging are usually roads, yarding, landings, and skid trails.

Roads: SmartWood wants landowners to assess the quality of all the roads on their property, and reduce negative impacts as much as feasible, which means accounting for the drainage of water at all points along a road, while assuming the worst (Photo 17). Once again, when managing in the present for the present and the future, one must consider how one's roads will function through time, which means taking certain precautions (Photo 18).

Well-built roads, for example, represent a good investment because they fail less often in an erosional sense, are easier and cheaper to

Photo 17 **Although the state of California has the most stringent forestry practice laws in the U.S., people ignore the regulations and the state has poor oversight, as demonstrated by this logging road in Mendocino County without waterbars to control soil erosion. (Photograph by Walter Smith.)**

maintain, and are available for all forms of forest access. It is nonetheless wise to severely restrict the use of roads and provide active maintenance during the wet season, in addition to which roads should be regularly monitored and repaired in a timely fashion.

In this sense, it is important to classify existing roads as permanent, those that should be closed, or those that should be removed. Old roadways can be useful, however, as trails or space for growing trees, in addition to which they can provide corridors for the falling of trees and access for skidding the logs, but they should never be left as active or potential sources of erosion.

Photo 18 A well-designed permanent forest access road in Mendocino County, California, with a rocked surface. (Photograph by Walter Smith.)

One must also be careful not to allow a road to divert a stream, and one must minimize the temptation to permanently fill small gullies instead of creating seasonal crossings. Overall, outsloped roads are best, especially if they have well-rocked surfaces. (An outsloped road is one that slopes away from the hillside so water will drain off of it without the need of a ditch.) On the other hand, rolling dips in roads are preferred to ditches situated where the uphill slope and road meet, which must then discharge their water through culverts placed under the road.

When culvert placement is necessary, assume that the next 100-year flood will occur during the first winter in which the new culvert is in place and use that assumption as the minimum standard when considering the potential loss of soil along the culvert's bottom. Finally, do not let concentrated drainage of water empty onto the nose of the ridge.

Yarding: Although logging clearly affects the land in a number of different ways, yarding (moving logs from the stump to the truck) is the activity most people associate with problems caused by logging. Yarding is difficult work and can be more environmentally damaging when trees are large, slopes are steep, and inexperienced operators are working in the forest than when trees are small or medium in size, slopes are gentle, and good, well-seasoned operators are at work.

SmartWood recognizes, however, that the way a site looks following logging is not necessarily a reflection of the degree of environmental damage. Slash (woody debris left on the ground following logging), bare soil, and gaps in the canopy are often useful elements in forest management and must be considered as active components of the recovering forest stand.

Yarding practices that are not suitable for a site can produce changes in topography and drainage, areas of bare soil oriented downhill, and gaps in the canopy larger than needed for regeneration. Signs of poor yarding are deep cuts by tractors, logs dragged through watercourses, deep deposits of slash, and/or mounds of bare soil.

Landings: The placement of landings, or the selective reopening of landings, is a critical choice for the forest practitioner. While landings should be minimized on a harvest unit, they must be adequate in both number and placement. Too many landings, or landings that are too large, remove productive growing space for trees and unnecessarily disrupt the forest ecosystem. On the other hand, too few landings means that logs must be dragged further over difficult terrain, creating more adverse effects on soil and water than is necessary.

Skid trails: Use cable yarding where feasible, which minimizes damage to the residual stand (remaining live trees) and other forest resources during harvest. Where skidders are used, establish a permanent system of skid trails. The total layout of a given harvest unit should be so conceived and implemented as to minimize disturbance to the site and thus the potential for soil erosion. Skid trails should be selected and marked on the ground prior to operations. Trees should not only be felled toward skid trails and corridors but also be bucked (have the limbs removed and cut into proper lengths) prior to skidding. Do not place skid trails in areas that contain watercourse zones. Finally, all the above dos and don'ts lead us to the topic of loggers.

There is a tendency to lump loggers into one stereotype and blame them for all the problems that arise from cutting trees, such as those discussed above. Loggers as a group, however, are given a bad rap. They are not the scourge of the forest as they are often made out to be when one looks for a scapegoat.

First, it must be remembered that loggers are people and some are more conscious of the consequences caused by their actions than are others. Second, "The American public for many years has been abusing the wasteful lumberman. A public which lives in wooden houses should be careful about throwing stones at lumbermen, even wasteful ones," admonished Aldo Leopold in 1928, "until it has learned how its own arbitrary demands as to kinds and qualities of lumber, help cause the waste which it decries." Be that as it may, any forester or landowner will tell you that the logger is the essential element to successfully implementing environmentally sound silvicultural practices, particularly when dealing with systems of selective logging.

One particular logger I (Walter) know goes to extraordinary lengths to ensure that the silvicultural plan is well executed and that forest values are protected. He and his crew go beyond the minimum expectations of both the management plan and the forest manager by eliminating the use of skid trails where possible, pulling cable to winch logs to the skidding equipment instead of moving the equipment off skid trails, putting protective bumper logs against trees that are to remain after harvest, and roping boards around trees on the landing to keep them from damage, all the while looking out for habitat and other ecological features that need protecting.

In addition, the logger and the forester must always have a good two-way communication. When the logger is in doubt about something, he waits for consultation with the forester before proceeding. When I was auditing one certified resource manager that this logger and his crew work for, the logging supervisor summoned the forester to talk about an area that the logger felt was too sensitive to cut. It was a very steep area near a stream that may have been overlooked by the forester. The logger wanted to eliminate this area from the harvest unit. After an on-the-ground review, the forester also determined it to be ill advised to harvest in that part of the unit and eliminated it.

As one can imagine, the reputation of this logger and his crew is excellent with landowners, and their services are sought by many foresters. The logger has, however, chosen to work exclusively with one certified resource manager for a number of years. In fact, the existence of a true logger–forester team is probably one of the most consistent characteristics of certified forest operations.

Many people may believe this to be an unusual situation, but I have found it quite common among loggers who work for resource managers and landowners who are certified. The key to good logging is intelligent and caring loggers who pride themselves in doing excellent work. Additionally, long-term, even multi-generational involvement with one forest management company or landowner gives the logger a sense of both ownership and stewardship of the land on which he works.

As previously stated, consistent and balanced attention to environmental, silvicultural, and socioeconomic conditions and priorities must be visible in all operations, ranging from protection of riparian buffers or wildlife nesting trees, to reducing residual stand damage, increasing the amount of coarse woody debris during timber harvesting, and the handling of chemicals or other materials. The following criteria focus on how carefully and consistently the forest management plan and management practices are implemented.

Sustainable Yield

> **4.1 — An annual or periodic cut has been established by area or volume, based on conservative and well-documented estimates of growth and yield, to ensure that the rate of harvest does not exceed the sustainable rate of growth. [If the yield is biologically sustainable, it can also be *biologically sustained.*]**

The objective of most certified forest owners and managers is to be conservative with their estimates of growth and yield, thereby cutting less volume than they grow in order to increase the available timber inventory in the future, which means growing trees in such a way as to increase their height and diameter, as well as the size and closure of their canopies. These owners and managers generally cut 2.5 percent or less of the standing inventory on an annual basis. They combine this strategy with a silvicultural prescription of thinning from below. Only a few landowners are fortunate enough to start with an uncut forest; it is thus imperative for them to restore the forest to a more productive state and ecological balance.

> **4.2 — The proposed annual of periodic cut (or other harvest calculation) is being followed in the forest.**

> **4.3 — Growth rates, stocking, and regeneration are being monitored by a suitable continuous forest inventory system.**

> **4.4 — Post-logging assessments take place to evaluate the effect of logging on future crop trees and general forest conditions.**

Photo 19   Note the old tree left in this thinned stand that will one day become a large fallen tree. As large, woody material on the ground, it will act not only as wildlife habitat but also as a reinvestment of biological capital (organic material) in the forest soil. (Photograph by Walter Smith.)

Silviculture

4.5 — Rationale behind the selected silvicultural prescription(s) is appropriately based on site-specific field data, local ecological conditions and disturbance regimes, local experience, and/or the published results of research pertinent to the forest in question (Photo 19).

4.6 — Silvicultural prescriptions (pre-, during, and post-logging) are being adhered to.

4.7 — Openings in the canopy are sufficient in size to regenerate the tree species of interest, but small enough to minimize unnecessary disturbance to the species composition, structure, and function of the ecological feedback loops in the forest (Photo 20).

4.8 — Sufficient action is taken to ensure the quality and quantity of timber in future stands (e.g., there is adequate natural

Photo 20 Creating openings in the forest canopy with selection cutting in Mendocino County, California. (Photograph by Walter Smith.)

Photo 21 Structurally superior tree in the Philippines marked to be retained as a possible source of seed. (Photograph by Walter Smith.)

regeneration and/or plant trees, an adequate number of seed trees are left, and residual trees are protected) (Photo 21).

4.9 — Seed stock is obtained from appropriate seed zones and, where feasible, is collected from trees growing naturally on or near the site in which they are to be sown as seedlings.

4.10 — Timber management protects the forest from the over-cutting of individual tree species or shifts in species dominance, unless stands are consciously being restored to an appropriate ecological condition.

4.11 — Timber management protects against the erosion of genetic material (genotype or phenotype) as a result of preferentially cutting the most desirable trees in terms of their health and shape.

Road Design, Construction, and Maintenance

4.12 — Standards for the placement, construction, maintenance, and closure of roads and skid trails are established to minimize environmental impacts. These standards are applied in the field.

4.13 — The system of roads and skid trails is kept to the minimum density necessary to provide access for management, is laid out according to topographic feature, and designed to use old roads where feasible.

Roads can be one of the most environmentally damaging aspects of forest management. Two forest owners in Northern California have built superior all-weather roads to both minimize soil erosion and create excellent forest access. Permanent all-season roads are outsloped, heavily rocked, and constructed to withstand traffic from logging trucks over the long term. The roads are also gated to keep unwanted and unnecessary traffic off them.

4.14 — Road fill is kept out of stream courses.

4.15 — Road drainage is designed to minimize surface erosion of soil, reduce ponding, and to protect stream integrity at crossings.

The owner of a small acreage in the northern Willamette Valley of Oregon is in the process of permanently outsloping all of the logging roads on his property. It will take him a while because of the expense, but he is committed to minimizing surface erosion, eliminating the need for ditches and culverts, both of which will minimize soil erosion during the long, wet winters and thereby protect the streams on his property.

4.16 — Culverts in fish-bearing streams are designed and installed to allow fish passage during low and high flows and to accommodate 100-year flood events (spanning bridges, oversized culvert bridges, and/or temporary structures may also be used at new sites).

A timber company in northern California not only pulled the road culverts where two streams joined each other but also replaced them with two flatbed railroad cars, which made adjoining bridges that can well withstand a 100-year flood event. It is one of the best arrangements that I (Chris) have seen, short of pulling culverts and permanently decommissioning roads.

> **4.17 — Road closures (taking out culverts, installing no maintenance structures, blocking road entrances, stabilizing old fill and crossings, outsloping surface, and so on) are implemented when and where necessary.**

Tree Felling

> **4.18 — Directional felling is used to minimize damage to residual trees, decrease the distances that require skidding, improve efficiency, and protect future trees retained for harvest as sources of seed.**

In almost every case, certified forest landowners and managers require directional felling to reduce impact on the forest and the remaining trees (Photo 22). Because the stand of trees left after logging represents future income, there are potentially large economic losses in the form of damaged trees that not only fail to meet their growth potential but also fail to meet the higher grades of lumber due to the damage they sustained during logging. In addition, excessive damage from logging can have a noticeably negative effect on the ecological integrity of the remaining forest.

Fallers for these forestland owners and managers know how to make felling cuts that direct the tree to the desired location. Undercuts are wide, cleanly cut, and accurately aimed (Photo 23). The fallers carry wedges for felling trees in the opposite direction of the tree's lean if that is the desired direction of felling (Photo 24). Sometimes fallers even carry hydraulic jacks with them so they can push a heavily leaning tree in the oppose direction of its lean if that is the way they want it to fall (Photo 25). Additionally, good loggers also pull trees with logging equipment to ensure that they can fell a tree in an appropriate direction.

> **4.19 — The rationale behind the selection of trees for cutting is transferred to the logger by marking the stand prior to cutting and/or adequate training and on-site supervision.**

> **4.20 — Areas of highly erosive soil, soil that is permanently saturated, critical riparian areas, and so on, are protected**

Photo 22 Directional felling means to aim a tree to fall in a predetermined place. (Photograph by Walter Smith.)

from timber cutting unless justified by sound ecological rationale and necessity.

4.21 — Tree felling does not occur on steep slopes (>75 percent or 35 degrees) unless experience, site conditions, and equipment can justify such an action by reducing negative environmental impacts.

Felling (cutting a tree down) is the beginning step in the actual harvesting process. If the felling job is poor, felled trees are simply crisscrossed and hung up in other trees, rather than being cut so as to fall toward the skid trail or yarder corridor. If great care is not taken

Photo 23   Aiming a tree to fall in a predetermined place is especially important in selective logging, where it is cost effective to minimize the impact on the forest and remaining trees. (Photograph by Walter Smith.)

during the actual cutting of trees, the job will be more expensive and the environmental damage will be greater. Logs that are crossed will knock down small trees when pulled to the skid trail. Additional roads may be necessary if the logs are not reachable from the existing road system. Most of the loggers that work for certified operations require directional felling and the fellers carry the appropriate equipment, such as wedges or hydraulic jacks. At times, a tree is pulled in the right direction by having a climber put a cable in the tree that is also attached to a tractor's winch. The feller then makes the appropriate cuts, and the tractor operator winches over the tree.

Skidding and Yarding

> 4.22 — **Routes for yarding and skidding are designed and located prior to harvest.**

> 4.23 — **Skid trails and yarding systems are laid out in such a way as to minimize damage to the residual stand, other forest resources, and the potential for soil erosion.**

Photo 24 A feller using wedges to make a tree fall in the selected place. Note the hard hat and face shield. (Photograph by Walter Smith.)

In the Willamette Valley of Oregon, there is a small forestland owner who is so concerned about protecting his land that he has his forester permanently lay out skid trails in such a way as to minimize, to the greatest extent possible, their impact on the land. This is not to say that the pattern of skid trails was inflexible. One skid trail, for example, is not where the landowner wanted it because a wildlife tree is currently in the way. When the tree falls, which will be of its own accord, and decomposed into the soil, the skid trail will be moved — but not before.

4.24 — Operations occur only on firm soil, not when soil conditions would cause severe rutting, compaction, or disturbance.

Photo 25   A feller using hydraulic jacks to fall a tree in the opposite direction of its lean. (Photograph by Walter Smith.)

4.25 — Log landings and cable yarding corridors are kept as small as possible; size and placement are determined by existing ecological conditions and logger safety.

One certified resource manager has made a conscience decision to use skyline logging on slopes over 40% even though there are old skidtrails on those steeper slopes. In some cases he is rehabilitating the old skidtrails; in other cases they are revegetating on their own. The yarder corridors are only 8 to 10 feet wide and are incorporated into the cutting plan, as are designated bumper trees. He only does selective harvesting on the forests that he manages and uses the skyline system successfully. With the narrowness of the corridors, they are hardly discernible when harvesting is complete. Again, the up-front economic costs are seen collectively as a long-term investment in the biological productivity of the forest by eliminating skidtrails that compact soil and take up growing space.

5 — Environment

The environmental criteria assess the effect of the applicant's management on the forest ecosystem and surrounding landscape.

5.0 — Environmental Impacts and Biological Conservation

Certification requires that forest managers place great attention on the protection of healthy ecosystems and the protection and restoration of endangered ones (e.g., wetlands or old-growth forest, Photo 26), conservation of threatened/endangered species, and precautionary use of chemicals. This care and protection is important because forests are far more than merely suppliers of wood fiber. They are the main source of water for most of the people of the world, and the water they produce and

Photo 26   Interior view of a healthy old-growth Douglas fir forest. (USDA Forest Service photograph.)

Photo 27   Deep snow stored in high-elevation forests is the main source of water for much of the world. (Photograph by Chris Maser.)

store greatly exceeds the value of whatever wood fiber humans may glean from them (Photo 27). They supply habitat for insects, birds, and bats that pollinate crops and for birds and bats that eat insects considered to be harmful to people's economic interests, such as the forest trees themselves. Forests are a major source of the oxygen we breathe, which has no substitute; all the while the trees store within their wooden bodies vast amounts of carbon, which helps to stabilize the global climate.

Environmental protection and biological conservation in the management of certified forests therefore include a combination of protective and restorative measures. Protective measures will focus on ensuring that all staff and contractors are cognizant of sensitive areas and take actions to protect them. Restorative measures may include efforts to increase the overall ecological integrity of the certified forest itself and the greater landscape surrounding the forest. Having said this, it must be understood that any stands of trees grown or manipulated after certification has been awarded by the Institue of Sustainable Forestry/SmartWood do not and will not qualify as ancient forest. We say this because such trees either are too young to qualify as ancient or will have lost their ecological integrity as an ancient forest ecosystem.

The following criteria are aimed at ensuring that environmentally sound protective and restorative measures are applied as needed in the certified forest to maintain and/or enhance its social–environmental sustainability over time. In discussing these criteria, I (Chris) will include examples in which people are concerned enough with the health and welfare of their land that they have, on their own, gone beyond the normal behavior of most forestland owners in protecting their land for the next generation.

Biological Conservation

5.1 — Field assessments of biological resources have been conducted.

Although I cannot recall a forestland owner with whom I have worked who had completed a sophisticated assessment of biological resources on his land, other than a timber inventory, prior to applying for certification, there is one interesting example of foresters who were astute enough to assess a potential danger to an important nontimber resource and work to correct it.

Tribal Forestry from an Indian reservation in northern California worked with the U.S. Forest Service in an attempt to revise the latter's management plan so that raking the duff layer off the forest floor was outlawed in the taking of certain mushrooms, because such raking was exceedingly detrimental to forest floor. Tribal Forestry asked the U.S. Forest Service to allow mushroom gathering only when the mushrooms were large enough to be positively identified and picked without disturbing the duff layer of the forest floor, which is how they managed the same mushroom resource within the confines of the reservation.

5.2 — Environmental impact assessments have been completed prior to commencement of activities in the forest.

5.3 — A consistent, scientifically sound monitoring system is accurately implemented and used to make adjustments in management that are commensurate with the intent and legal mandates of biological conservation.

Although most small forestland owners with whom I have worked have had little notion of what a scientifically sound monitoring system is, a few of them were so observant that they eagerly told me how the deer used their property, where the red fox den was, about the bear tracks they saw along a ridge top early last summer, and so on. While these observations may seem random and of little consequence, they not only tell me much about the landowner's attitude toward his or her land but also that the landowner is interested enough to willingly follow more systematic and scientific procedures in monitoring the effects of his or her forest practices once he or she understands why monitoring is important and how to go about it. This willingness to take the next step, to elevate one's level of consciousness, is critical to the spread of certification, which is, after all, built on an attitude of voluntarily going beyond the norm in behavior as a trustee of one's land for those who must live on it when the current owner is no longer around.

5.4 — Management activities not only consider but also actively integrate landscape-level concerns, such as landscape-scale diversity, on neighboring properties.

5.5 — The forest contains a mixed composition of species that creates a physical structure with multiple age and size classes of trees, including a proportionately significant amount of late succession forest.

The best example I have seen of a forest that fulfills this criterion was that of an Indian tribe on tribal lands in northern California. The people in Tribal Forestry, through their forest management plan, were so sensitive to and concerned with protecting and maintaining some of their cultural ways that they put areas of ancient forest off limits to any commercial use because they were of greater value for spiritual reasons than for timber.

Other areas of old-growth Douglas fir were carefully managed because, by tribal custom, the red top-knots of the pileated woodpecker and the fur of the fisher were both used for ceremonial purposes (Photos 28 to 30). By following the spirit and the intent of the forest management plan, Tribal Forestry was not only protecting the habitat of those species with cultural importance but also their own culture and that of the people whose land they were charged to care for. The foresters knew that, and they honored it, despite the foregone revenue and the sometimes contentious nature of the Tribal Council, which at times seemed more interested in the immediate revenue than in protecting tribal customs.

The forest management plan, written by the people in Tribal Forestry, also directed the protection and maintenance of areas of bear grass, which the women used for making baskets, and areas of hazel bushes, the nuts of which the people gathered to eat. In addition, specific, traditional areas of tanoak were purposefully managed to protect and maintain potential crops of acorns, which the people used as food.

Although the foresters designed small openings in the forest, they left islands with a mix of live trees as habitat for wildlife, in addition to a

Photo 28 **A mix of species and age classes in an old-growth forest. The tree in the immediate central foreground is a western redcedar. The tree just to the right and behind the cedar is a Douglas fir, whereas the tree to the left of the cedar in the fore area of the photograph is a western hemlock. Each species of tree has different attributes with respect to its functionality as habitat for wildlife and how each lives, grows, dies, and recycles into the soil because each species differs in its structural characteristics and chemical composition. (Photograph by Chris Maser and Larry D. Harris.)**

scattering of available snags and large fallen trees on the ground. When they made shelter-wood cuts, the remaining overstory was off limits to further harvest because it was needed to maintain the large trees in order to create not only the characteristics of an all-age forest as the young trees began to grow but also as a legacy for future snags and coarse woody debris on the ground. This was all done in complience with the tribal forestry management plan.

5.6 — Endemic, threatened, rare, or endangered species or ecosystems (on either local and/or international listings) are explicitly protected during forest operations.

One small woodland owner had a spring on his land that came out of a rocky hillside at the head of a little ravine, which supported large, old Douglas fir trees, as well as big-leafed maples and other species.

Photo 29   A large old-growth Douglas fir, weakened by root-rot fungus, and toppled by a strong wind. The old tree is momentarily hung up in a neighboring tree, but will eventually add its structure to the forest floor and its body as a reinvestment of organic material and elements to the soil and, through the atomic interchange of the soil, to the forest of the future. (USDA Forest Service photograph by James M. Trappe.)

Having once seen a salamander by the spring, the owner put the entire ravine off limits to any timber harvest or other forestry practices because, while he did not know what kind of salamander he had seen, he thought it might be an endangered species.

When I pointed out that the salamander was unlikely to be an endangered species, based on the geographical location of the spring, the landowner nevertheless decided to maintain the ravine's protected status because it made him "feel better to be on the safe side."

5.7 — Conservation areas are configured at an ecologically and spatially appropriate scale, clearly outlined on maps and demarcated in the field, and forest operations are carefully controlled in these areas.

This criterion is more difficult than most to fulfill with any degree of certainty on small acreages because that which constitutes an ecologically and spatially "appropriate" scale may, in fact, require an area far larger

Photo 30 A fallen old-growth tree and a fallen old-growth snag in a high-elevation forest, where they add different kinds of structure and function to the forest floor. Over time, recycling wood can be found virtually throughout the upper soil layers of old-growth forest, where it acts as a legacy of soil fertility and health from one forest to the next. (Photograph by Chris Maser.)

than the person's holdings. Appropriate scale, in this sense, refers to the landowner's attitude and willingness to designate the conservation areas on his or her property to the very best of his or her ability, given the limitations of how the areas in need of protection are configured across the ownership and how much of the ownership they encompass.

> **5.8 — Areas of ancient forest (stands of 10 acres or more, where at least 70 percent of the basal area is composed of trees 150 years old and older) are identified, mapped or located on aerial photographs, and protected from cutting.**

See example under Criterion 5.5.

> **5.9 — Where appropriate, fire is reintroduced to counter the long-term ecological effects of fire exclusion. It is understood, however, that such reintroduction of fire is subject to**

restrictions regarding safety, timing, intensity, liability, and so on.

Referring again to Criterion 5.5, Tribal Forestry, following the guidelines in the tribal forestry management plan, pursued a policy of annual or biannual burns in the areas that had bear grass, hazel bushes, and tanoak trees, much as their ancestors had done, in an effort to maintain the sustainability of the nonforest products that were important to tribal members.

5.10 — Use of exotic species is discouraged. If an exotic species is used for a well-justified and specific purpose, however, its use is carefully controlled and the outcome is monitored for potential negative environmental impacts.

I have yet to find a small woodland owner, who has applied for certification, to have knowingly planted exotic species on his or her property. It does not seem to be something they would think of doing. Having said this, Richard Donovan, director of the SmartWood Program, informs me that, outside of the Pacific Northwest, some people do purposefully introduce exotic species.

5.11 — Use of biological control agents is documented, minimized, monitored, and strictly controlled in accordance with national laws and internationally accepted scientific protocol.

5.12 — Indigenous species are used to the exclusion of genetically engineered organisms (including trees) and invasive exotic plants.

(For large landowners [>1,000 acres].)

5.13 — Based on the identification of critical biological areas, representative samples of ecosystems that are lacking within the landscape are maintained and/or restored.

One landowner had a knoll on his property that supported a wonderful grove of Oregon white oak, a species that is rapidly disappearing as a viable habitat in the Willamette Valley of Oregon. Recognizing this, the landowner set the whole hill aside to protect and maintain the oak

Photo 31   A female rufus hummingbird at her nest in an Oregon white oak tree. The nest is camouflaged with lichens from the oak. (Photograph by Chris Maser.)

community. Oregon white oak is one of the habitats used for nesting by the rufus hummingbird (Photos 31 and 32).

> **5.14 — Harvesting maintains at least 85 percent of the property in forest stands *older* than 10 years with good canopies, which means that up to 15 percent of the property can be in forest clearings or timber stands less than 10 years old.**

Stream Protection

> **5.15 — Riparian areas are managed to maintain water quality and temperature, clean spawning gravel, and protect the integrity of stream structure (i.e., pools favored by anadromous fish).**

A large timber company, of its own accord, exceeded the legal requirements to protect streamside habitat in that it not only extended the buffer zone well beyond the mandated area but also made them a permanent part of habitat connectivity on their property, which amounted to a lot of acreage being taken out of the land base for commercial harvest. Such

Photo 32   Same hummingbird nest as in Photo 31. Note the baby and the unhatched egg in the nest. (Photograph by Chris Maser.)

protected streamsides are a major source of large wood debris that on the one hand protects a stream's bank, and on the other forms habitat for fish, such as pools (Photos 33 and 34).

5.16 — Wetland areas are protected and restored to maintain ecosystem functions that support wildlife, moderate stream flow, and improve water quality, flood control, trapping of sediments, and so on.

I remember one landowner, for example, who discovered that he had a beaver move into a small trickle of water and build a dam. Over a couple of years, the dam created a small pond and the beginnings of a surrounding marshy area. Noticing that the dam not only stored water in winter but also released water in summer that made the once-intermittent trickle into a permanent trickle, the landowner drew a line around the dam and put it off limits to timber harvest or other forestry activities.

When I told him that beaver dams alter a stream's channel not only by impounding water but also by giving the channel gradient a stair-step profile, which decreases the velocity of the current, expands the area of flooded soils, and increases the retention of sediments and organic matter,

Photo 33   Coarse woody debris in a stream that is flowing away from the viewer. Note the fallen tree at four o'clock, which is snuggled close to the stream's bank and thus protects it from erosion by the current. Note also the large rootwad in the left side of the photograph, which is anchoring the two recently fallen trees to the bank of the stream. The fallen tree in the middle of the stream is providing habitat for fish on its downstream side, while at the same time protecting the stream's bank as seen in the lower right-hand corner of the photograph. (USDA Forest Service photograph.)

he was tickled and asked me to tell him more. So I told him that the wetted area would in turn increase the beaver's supply of food and offer protection from predators. In addition, the wetted area would store nitrogen, which is converted to nitrate and becomes available for uptake by plants when dams either break or are abandoned by the beaver, which allows the surrounding soil to be exposed to air.

I went on to explain that beaver impoundments retain other important chemicals, like potassium, calcium, magnesium, iron, and sulfate, which become nutrients when in the correct proportions and available to plants when dams either break and/or ponds are deserted by the beaver.

Beaver also influence the community dynamics of riparian vegetation, instream bottom-dwelling organisms, fish, wildlife, the structural diversity of streams, and the nutritional value of certain species of trees. When a beaver pond drains after the dam has been abandoned, small grassy areas

Photo 34   A large, stable fallen tree that is holding back sediment, stabilizing the stream's bank, dissipating part of the stream's energy, and creating a waterfall and its attendant pool, which is excellent habitat for small fish. (USDA Forest Service photograph.)

or even meadows form on the newly exposed sediment, which can be quite stable, unless, of course, the beaver returns and repairs the dam, once again flooding the area.

Other ponds gradually fill in forming wet meadows, even with a resident beaver in charge. This happens because the calm water of a pond allows suspended sediments to settle out — a major factor in building up soil that eventually becomes a meadow. A dam and pond have the additional benefit of capturing eroded soil close to the source of erosion rather than allowing it to wash downstream, where it causes turbidity in the water as it descends the stream.

Beaver thus create structural diversity not only directly by felling trees and building dams but also secondarily by flooding areas that kill trees (as was the case in this dam, where one or two trees were beginning to die), which in turn creates standing dead trees or snags that are important to numerous insects, birds, and mammals. And beaver even affect the chemical composition of some trees.

In addition, a beaver dam can have a profound effect on the riparian habitat surrounding a pond, by dramatically altering both how water flows and is stored in a stream's immediate drainage way. A beaver dam and the pond it creates cause more water to be held in the subsurface soils surrounding the dam and pond than would exist without them.

Subsurface flow of water caused by a beaver dam and its associated pond contributes to the general diversity and richness of species along a specific area of the stream. Such water is particularly important to species adapted to or requiring wetlands. In addition, the subsurface flow of water from the beaver pond around the dam and back into the stream below the dam cools the water in summer and warms it in winter.

It works like this: in summer, as water from a beaver pond enters the subsurface flow, the temperature of the water is generally about as warm as that of the pond. As it moves slowly through the subsurface soil and ultimately returns to the stream below the dam after 2 or 3 months, it is cooler than it was when it first seeped out of the pond and into the soil. This means that water reentering the stream is relatively cooler than the stream itself, which represents a mechanism that creates locally cooler stream temperatures in the heat of summer.

The reverse is true in winter. Namely, the water reentering the stream after passing through subsurface soils for 2 or 3 months is warmer than when it initially seeped from the beaver pond into the soil.

When I finished my explanation, he looked at me and said that he would never have guessed a beaver could do all of that. In response to what he had learned, he outlined a bigger conservation area on his property to include a potential wetland below the beaver's dam and a buffer around the dam and potential wetland to provide food and protection for the beaver.

5.17 — Forest managers exercise maximum caution to maintain or restore ecological integrity within an inner buffer zone of 50 to 75 feet (minimum width) on both sides of the active high-flow stream channel unless another width is scientifically justified for the site (Photos 35A and 35B).

The same timber company discussed in Criterion 5.15 worked with its fisheries biologist to restore both the health of the riparian area and the health of the stream in one particular part of their land holding. This work exemplified the spirit of both state and federal laws enacted to protect the health of streams for fish and wildlife, as well as water quality.

5.18 — Forest managers exercise caution to maintain or restore ecological integrity (but may operate more freely) within an outer buffer zone of 75 to 150 feet (minimum width) on both

Photos 35 A and B Certified forest operations that have protected trees in the riparian zone to fall into the stream and thereby add structural and functional habitat that not only benefits fish and other animals but also helps to stabilize the stream. (Photographs by Walter Smith.)

sides of the active high-flow stream channel unless another width is scientifically justified for the site.

5.19 — Management activities prevent surface erosion of soil and landslides, and reduce the impact of peak water flow to riparian and aquatic habitats.

Retention of Critical Forest Structure

5.20 — Forest managers retain and/or recruit sufficient large woody material on the ground to protect species of indigenous fungi, plants, and animals, soil fertility and integrity, and overall ecosystem functions (Photos 36 and 37).

A timber company developed the policy of so carefully checking logs before they were yarded to the landing that, relatively speaking, few of them were hauled off site. When cull logs did inadvertently end up on the landing, they were, whenever feasible, taken back to the area from which they came and left on site as part of the biological reinvestment in the soil.

5.21 — Forest managers retain and/or recruit sufficient large standing dead trees (snags) to protect species of indigenous fungi, plants, and animals and the overall integrity of ecosystem functions (Photo 38).

Although the following story is not about a person whose forest has been certified, it exemplifies the kind of open-minded attitude found in many of the people who either own certified forests or who manage them. And it is precisely one's positive attitude, willingness to learn, and willingness to teach by sharing what one has learned that is one of the real values of forest certification.

In 1978, I was asked by a district ranger if I would visit his district, in the Wenatchee National Forest in the state of Washington, where he wanted me to present a workshop on the role of dead wood in the forest ecosystem. When I arrive, however, the ranger was nowhere to be found. Instead, he had assigned his young timber management officer to the position of host.

After spending a day in the conference room viewing slides of standing dead trees or snags and fallen trees and discussing how they functioned as habitat and sites of nutrient cycling in a forest, we all (around 30 people) went into the field. One of the many stops was a stand of large, old ponderosa pine, which had an old, Forest Service-style telephone line

Photo 36   A precommercial thin in which the certified operator lopped and left woody material as a reinvestment of biological capital in the forest soil. (Photograph by Walter Smith.)

extending through part of it. The pines themselves were widely spaced with trunks free of branches for 50 feet or so — a real park-like stand.

After walking through the stand for about half an hour while the timber management officer explained how cutting this particular stand was part of a watershed management plan, I asked the entire group to stop, be silent, and listen.

After some minutes of silence, I asked, "What do you hear?"

"Nothing," came the general reply.

"Exactly," I said, "there are no diseased, dying, or dead standing trees, and there are no large dead trees on the ground. In other words, the

Photo 37   A log left by a certified operator as a reinvestment of biological capital in the forest soil. (Photograph by Walter Smith.)

habitat is sterile, but if you look at the telephone poles, you will find a woodpecker nest cavity in every one."

At this point, the timber management officer asked me what I would advise.

"The ball is in your court," I said, and asked him what he thought ought to be done, based on the slide presentations and discussions, whereupon he went to a vehicle and returned with a small ax.

Handing me the ax, he said, "Chris, begin unmarking some of the trees to leave standing for snags and logs in the future."

I looked at him, handed him the ax, and said, "It's your forest, your management plan, and you're in charge; you unmark the trees you think are the best to save for snags and logs."

Without a word, he took the ax and began obliterating the "to cut" marks on a goodly number of trees.

The upshot of this story is that by the early 1980s the timber management officer was a district ranger who was vitally interested in applying all the latest research findings on the ground. By the mid-1980s, he had been a ranger in two districts, both in Oregon, and went to Washington, D.C. By the 1990s he was a forest supervisor in Minnesota, and by 2000, he was a forest supervisor in California.

Photo 38   A snag (dead standing tree) left in a certified forestry operation that is used by cavity nesters; note the cavities up and down the snag. (Photograph by Walter Smith.)

I worked with him periodically over the years and gained a good sense of just how much he taught by example to the people with whom he worked. As district ranger and forest supervisor, he left a legacy of enlightenment everywhere he went. In addition, he was constantly involved with citizen groups with whom he was always enthusiastic, always willing to learn, and always teaching by example.

Now, if one Forest Service employee can do that much good, how much more good can a growing number of people do who own forestlands on which excellent forestry is practiced? How much more good can certified forest managers do when they care for the forestlands of numerous clients? These people are, after all, constantly teaching by example.

5.22 — Forest managers retain and/or recruit sufficient large "legacy trees" (e.g., old growth) to protect species of indigenous fungi, plants, and animals and the overall integrity of ecosystem functions.

One forestland owner had a little difficulty understanding why I suggested that he tag with a metal numbered tag all the trees he intended to leave as legacy trees for use by wildlife, to become snags, and to fall as large down woody debris to recycle into the forest soil. After some discussion in the field, he suddenly understood the value of the historical aspects I was talking about.

As a result of his new understanding, he secured a numbered metal tag to every legacy tree on his entire property so he could keep records, both written and photographic, of what happened to them and how they were used as habitat over the years. The tags also reminded everyone that these trees were permanently off limits to logging no matter how good the market was.

5.23 — Hardwoods and understory vegetation are retained as necessary to maintain and/or restore a balanced mix of species, forest structure, and function over time.

The forest management plan drawn up by the Tribal Forestry staff of a northern California Indian tribe specified that hardwoods were to be maintained, as much as possible, in the historical mix of species because they were important to the overall health of the tribe's traditional forestland and the tribe's traditional use of that land.

Chemicals

The following discussion on chemicals is based on a conversation I (Chris) had with Dr. Steve Radosevich, a professor of Forest Science at Oregon State University who not only studies strategies of managing unwanted vegetation with and without herbicides but also has participated in a number of SmartWood and at least one Scientific Certifications Systems field assessment to review applicants' forest operations for possible certification. According to Dr. Radosevich, almost every assessment he has been part of deals with the issue of herbicide use in some way. Even the Institute for Sustainable Forestry, which says "Forest practitioners will not use artificial chemical fertilizers or synthetic chemical pesticides," makes their use conditional. The use of chemicals is therefore interpreted to mean that, while some use may be acceptable in the short term, the goal is to minimize their use or, preferably, to completely eliminate use of artificial chemicals in certified forests over time.

Although some chemicals — chlorinated hydrocarbons — have already been eliminated from use in certified forests by the Forest Stewardship Council, this can either be a very long list of chemicals or a reasonably short one, depending on how chlorinated hydrocarbon is defined. Although there currently is a debate centered on the definition of chlorinated hydrocarbons, SmartWood has adopted the strict definition, which leaves open the possibility of using other pesticides.

The following questions are those that Radosevich asks during a field assessment in order to gain a perspective on the use of chemicals in forestry operations:

1. Does a clear-cut/herbicide treadmill appear to exist? If so, the use of herbicides is a symptom of a more deeply seated management problem.

2. Are herbicides being used for purposes of ecological restoration? If so, their use must be restricted to that use only — and then on a one-time basis. For example, where coastal redwoods have been clear-cut in northwestern California, which has allowed tanoak to take over vast areas, a nonchlorinated hydrocarbon herbicide could be used one time to help reconvert a given site back to redwoods. Another one-time use of a nonchlorinated hydrocarbon herbicide might be following a significantly large wildfire in a mixed conifer–hardwood forest to prevent stump-sprouting hardwoods from dominating the site because of the fire.

3. Does success with herbicide use for restoration carry over into other management? If it does, this is equivalent to the clear-cut/herbicide treadmill.

4. How can the use of herbicides be reduced and/or eliminated over time? In this case, careful monitoring is required to assure that the actual growth of unwanted vegetation really warrants chemical treatment. If individual data points are used rather than means in interpreting results from monitoring and/or experiments, chemical use can usually be reduced by 50 to 70 percent.

5. If herbicides are used, are they applied to edible foods of indigenous peoples or of other people who enjoy gathering such things as huckleberries and blueberries in the wild?

6. Finally, there must be no aerial applications of pesticides of any kind. All pesticide applications must be kept strictly away from any water, and no persistent or extremely toxic materials are to be used. (The Forest Stewardship Council needs to develop a list of these persistent and extremely toxic materials.)

5.24 — No synthetic fungicides, herbicides, and pesticides are used. If less environmentally hazardous methods have proven ineffective and the land manager has applied synthetic chemicals, plans are in place to phase out their use over the shortest reasonable period of time (Photo 39).

The artificial means of controlling vegetation and pests are generally not used or minimally used by certified forest managers. Several certified managers that I (Walter) have come in contact with have exemplary philosophies about these issues. First, they accept some loss from pests

Photo 39   Instead of using herbicides to control unwanted vegetation, a certified operator in Humboldt County, California, used hand thinning as a control. (Photograph by Walter Smith.)

and pathogens as a natural occurrence of forest ecology. Second, they use silviculture to manage the pest and pathogen problems that might occur, e.g., harvest lightly to maintain the diversity of forest species and structures so as not to disrupt the natural ability of the forest to combat these problems. Chemicals are generally used only in situations that require restoration, such as those in which the forest has been mismanaged in the past and has been overcome by pests, pathogens, and/or unwanted vegetation.

> **5.25 — The ban on certain chemicals in the U.S. and Europe, as well as World Health Organization Type 1A and 1B chemicals, is honored. In addition, no chlorinated hydrocarbon pesticides are used.**

> **5.26 — When chemicals are used, staff and contractors not only receive training in but also use the correct procedures when handling, applying, storing, and disposing of chemicals.**

> **5.27 — Only terrestrial application of synthetic chemicals is used.**

> **5.28 — Chemicals, containers, and inorganic wastes (both liquid and solid) are disposed of in an environmentally appropriate manner at off-site locations.**

Phase 5, Information from Other Parties

In phase 5, the team visits with people outside of the applicant's business in order to gain a sense of how the business is viewed by associates and the community at large. In this way, it is possible to assess the strengths and weaknesses of the business in terms of social relationships and to arrive at ways potential shortcomings might be remedied. The intent of this section, as in all the criteria, is to help the applicant improve his or her business, recognizing that support of a local business by local people is critical to the economic viability of the business.

To this end, meetings with the staff of local, regional, or national governments should — and in some cases *must* — do the following:

- Discuss legal or other issues related to the applicant's operation (*must*).
- Discuss the adherence of the applicant's company to forest regulations, payment (e.g., for stumpage), citation for violations, and so on (*must*).

- Discuss the existing or potential regional or national laws that might affect the applicant's operation or certification in the private sector.
- Discuss the status of forest certification in the region and/or country and get copies of any documents that relate to such certification.
- Review and/or get copies of the best management practices, applicable field-oriented documents concerning forestry practices, forest management guidelines, and laws or administrative decrees.
- Get and/or review lists of threatened or endangered species (*must*).
- Get and/or review lists of critical environmental organizations, community development organizations, or church groups that might have perspectives on the applicant's operation (*must*).
- Get and/or review lists of academicians, researchers, or other potentially important individuals.
- Get and/or review lists of people in affected communities.
- Get and/or review lists of chemicals that are banned or restricted from use in the country as relevant to the applicant's operation.

Meetings with critical local environmental, social, and community organizations or individuals should focus on the reputation of the applicant with respect to:

- Positive initiatives or conflicts related to the applicant's company, vis-à-vis social issues dealing with the welfare of employees and issues surrounding responsible citizenship
- Endangered and threatened species and banned or restricted chemicals (here it is preferable to obtain lists)
- The relations of the applicant's business with nongovernment organizations and other interested parties

6 — Social Issues

The section on social issues addresses the implications of the applicant's forest management on the employees, the local community, and indigenous peoples. (Indigenous peoples are considered in more detail under no. 9).

6.0 — Community and Worker Relations

Forestry actions are community actions. They affect neighboring landowners and workers on and off the site. They affect the well-being of the community by providing economic benefits (direct and indirect employment, taxes, and entrepreneurial opportunities), environmental benefits (wildlife habitat, clean water, clear air, ambient temperatures), and social benefits (recreation, spiritual renewal, and community ambiance).

Forest management for the long term involves maintaining good relationships with the local community. Landowners who conduct certified operations should be good neighbors, which may be fairly simple for relatively smaller operations, where neighborliness might entail such actions as responsible operation of harvesting equipment on local roads, protection of historic cultural or archaeological sites, or positive relationships with adjoining landowners.

For larger public or private operations, however, the implications are usually greater. Typically, such operations must give careful consideration to local recreational, environmental, and aesthetic concerns; provide opportunities for local employment; and utilize local services. Big operations should also develop a system for public involvement in the forest planning processes, especially when management activities encompass a large area of land. Clearly, public participation is crucial on publicly owned forestlands.

Beyond neighborliness, the quality of forestry achieved on the ground depends on the workers who plant the trees, fell the timber, build the roads, and so on. Developing a team of well-trained and motivated workers who are familiar with local forest conditions and the desired management style is an excellent investment. Local workers are often ideal candidates for hire because of their knowledge of local forest conditions, their relatively long-term availability, and their connection to the surrounding community or communities. Offering good forest workers continued job opportunities and paying them well serves the interests of workers, managers, landowners, community, and forest alike.

I (Walter) have had the privilege of assessing a company that really stands out in terms of community and worker relations. The company is a substantial wood manufacturing enterprise, around $600 million to $1 billion annually in sales. They own very little forestland, but what they do own is certified. A number of their manufacturing plants carry chain-of-custody certification. Their company vision encompasses a balanced approach of quality vs. quantity for the workers, community, environment, and business.

The company began in the 1950s with one manufacturing plant. In the 1970s, the owner began to sell shares of the company to the employees. In 1998, the principal owner sold his remaining shares and the employees acquired full ownership of the company, which now includes multiple plants in the U.S. and Canada and is still growing.

All employees participate in stock ownership, with company contributions to each employee's stock fund. The company contributes 15 percent of each full-time employee's annual salary to company stock after an employee has completed 2,000 hours of service. Employees are fully vested after 5 years of employment.

The company's hourly wages are clearly competitive and superior to existing wages in their area. They encourage and financially support employees who pursue continuing education by annually committing 40 hours to staff training and development for each employee. As one can imagine, turnover is extremely low, around 1 percent, and opportunities for advancement within the company are very good.

During the chain-of-custody assessment at one of the plants, a woman I interviewed in the production department said she had worked at the plant for 18 years, that she had started as a laborer and had worked at almost every job in the plant from janitor to forklift driver to machine operator. She knew the production process inside and out (which, by the way, was extremely helpful in our understanding of the chain-of-custody issues inside the plant) and eventually worked her way into the production tracking and accounting department. While it is not unusual for employees to advance to a managerial position in this company, it takes a long time because nobody quits.

Another incident that typifies this company was their response to an employee when he was faced with an $85,000 medical bill because of a family health-related catastrophe. The health insurance company refused to cover medical costs due to a snafu in the paperwork, so the company paid all the hospital bills.

The company treats their logging contractors the way that they treat their permanent employees. Managers stated a strong commitment to retaining good contractors. They pay above standard contract prices; contractors are paid by thousand board feet and tonnage. Not surprisingly, I found that local contractors consistently seek contracts with the company because remuneration exceeds local standards and the company strives to keep local contractors steadily employed. Additionally, much like their interest in certification of their forest management as a way of continually seeking improvements to their system and practices, the company has engaged a neutral, third-party firm to review company wages and benefits as well as contracts.

As a member of their community, the company got high marks from local citizens. Planned management activities are communicated effectively by word-of-mouth. Community members are often involved in the jobs, and are often present when jobs commence.

The company allows entry on their forestland for traditional community uses, which include hunting, harvesting nontimber forest products (such as mushrooms and material for floral arrangements), and the collection of firewood. A permit that outlines the individual's responsibility is required, however.

The company staff also participates in and supports a number of local activities that enhance community life, such as public schools, softball and little league baseball, two summer camps, a scholarship fund for local students, the Red Cross, a local wood chopping contest, and two volunteer conservation groups. Interviews with landowners, independent consulting foresters, staff from the Division of Forestry, and environmental groups indicate that the company is a responsible neighbor and community member.

Although not all certified companies go to these lengths, one feature that certified companies have in common is the basic philosophy that their employees and the local community are their biggest assets, and thus the main reason that they are successful.

In addition to local communities, there are Indigenous Americans within California who live on reservations. The policy of the Institute for Sustainable Forestry is to recognize the autonomy of Indigenous Americans within California. If the Institute for Sustainable Forestry/SmartWood guidelines for certification in California conflict with the laws and customs of Indigenous Americans, exceptions may apply. For management issues not addressed under tribal laws, the assessment team may grant exceptions to the California guidelines to accommodate contemporary and/or traditional uses that are approved by the tribe and implemented on its own property. At a minimum, however, all practices must comply with the international standards of the Forest Stewardship Council and the generic guidelines of SmartWood.

In order to certify lands under tribal ownership, assessment teams of the Institute for Sustainable Forestry/SmartWood must include at least one Indigenous American and all other assessors must have had previous work-related experience with tribal or indigenous operations. Confidentiality of disclosures shall be maintained in keeping with applicable laws and the desires of the tribal representatives.

This particular addendum is based largely on the philosophy behind Principle 3 of the Forest Stewardship Council, which states: "The legal and customary rights of indigenous peoples to own, use, and manage their lands, territories, and resources shall be recognized and respected."

The following criteria are designed to nudge all certified companies in the direction of sustainable forestry, be they in local communities or on tribal lands.

Community Relations

6.1 — Local contractors, workers, and companies are given preference in hiring with respect to forest management activities.

In the Willamette Valley of Oregon is a woman who manages certified forestland for a foreign owner. She has a personal relationship with the contractors and workers on the property, all of which are local. The contractors and workers clearly like and respect the manager and know they can count on having work whenever it is available. In turn, both contractors and workers not only take real pride in what they do and how they do it but also are open to learning, which on that piece of property is an ongoing process.

6.2 — Local communities or individuals affected by forestry activities are notified of proposed actions prior to their implementation and given opportunities to effect management decisions (Photos 40 and 41).

An Indian tribe in northern California was exemplary not only in its strategy to stimulate community involvement during the development of its forest management plan but also in the incorporation of feedback from the community into the elements of the final plan. Written summaries of the forest management alternatives (ranging from industrial-style management to low intensity caretaker management) and requests for

Photo 40 Community in the U.S. discussing forest management options. (Institute of Sustainable Forestry photograph.)

Photo 41   Workshop on certification in Japan. (Photograph courtesy of Walter Smith.)

comments were mailed to each tribal household early in the process, in addition to which public notices were posted. These efforts were relatively unsuccessful, however, in stimulating interest from the general tribal membership.

To address this problem, an exceedingly well-made video was recorded in which the alternatives being considered were outlined, and a copy was mailed to each tribal household. The video was very successful in stimulating community interest. Following dissemination of the video, a series of public hearings was convened with tribal staff; these meetings were not only well attended but also stimulated much discussion about public uses of the forest and the expectations of tribal members for its management.

The video was also successful in crystallizing the concept of land-use allocations for the Tribal Council (the strategic heart of the forest management plan). Public comment and debate helped stimulate the Council to further refine the middle-of-the-road preferred alternative to better meet the perceived needs of tribal members. After the Council approved the final alternative and before the management plan was finally approved by the Tribe and the Bureau of Indian Affairs, public comment was again solicited through the process established by the National Environmental Policy Act.

6.3 — The legal and traditional rights of local communities to lawfully use and manage forest resources (i.e., water, public rights of way, recreation, firewood, nontimber forest products) have been formally recognized where relevant, documented in written agreements if necessary, and honored through compliance with relevant legal procedures and covenants stemming from dispute resolution.

6.4 — Archaeological, cultural, historical, or religious sites of special significance are identified and protected.

6.5 — Compensation is provided to local communities for direct damage for which the forest operation is responsible.

6.6 — Managers participate with local organizations, educational programs, and/or state agencies to further public education, training, and research on forests and forest management.

6.7 — Forest management is benign to and protective of human health within the surrounding water catchments and communities.

6.8 — Impacts on the aesthetic character of the forest are considered in planning and accounted for when implementing management activities.

In northern California is a woman who owns about 40 acres of land, including magnificent redwood trees that abut a state park. The woman and her certified manager have examined the trees individually, and decided which trees will be allowed to age indefinitely and blend into the forest structure of the park and which trees will be cut to allow the legacy trees to grow. Over time, her forest land will take on the same forest structure as that of the park and appear inseparable, except for the inconspicuous wire fence that divides the two properties.

Worker Relations

6.9 — Wages and other benefits (health, retirement, worker's compensation, housing, food) for full-time staff and contractors are fair and meet or exceed prevailing local standards.

6.10 — The conditions under which people work are safe and meet or exceed local norms, which means that the number of accidents is also within local norms.

Health, safety, and welfare of the work force are vital elements in sustainable forestry. Workers that are fairly compensated and protected are the most likely to carry out silvicultural prescriptions properly and maintain the long-term relationship with the forest owner and/or manager so important to good management. Good loggers provide safety equipment and incentives for injury-free work.

One logger I (Walter) met shares the workman compensation cost savings with his workers if they remain injury free. The injury rate of his workers over the years is extremely low; thus, his workman's compensation costs are low and the workers return to their families each night as whole people. Of course, this logger also pays his workers above the industry standard and is rewarded with low personnel turnover and excellent rapport with the workers. He is also rewarded with lower personnel costs and better logging jobs.

6.11 — Safeguards exist to ensure that contractors and/or leasees comply with applicable labor laws.

6.12 — Workers are appropriately trained.

6.13 — Workers are given freedom to organize and negotiate with employers as per Conventions 87 and 98 of the International Labor Organization.

The Rights of Indigenous Peoples

6.14 — Openness and a spirit of cooperation and coordination are demonstrated in the planning and implementation of forestry practices in areas of importance to tribes of Indigenous Americans.

6.15 — An appropriate process for addressing and resolving grievances is in place with respect to disputes concerning the human rights of and traditional uses of resources by Indigenous Americans.

6.16 — Indigenous Americans are fairly compensated for the application of and/or commercial use of their traditional

knowledge regarding the use of species in the forest and/or management practices in forestry operations.

7 — Economic Viability

The criteria under economic viability are designed not only to assess the degree to which an applicant's operation can survive as a business and the adequacy of the reinvestment of biological capital in the forest ecosystem but also to assess the degree to which the value of the forest products is captured, the efficiency and effectiveness of their utilization, and the adequacy of their marketing.

7.0 — Benefits from the Forest and Economic Viability

Well-managed forestry operations must be able to harmonize the economic value of forestry with the ecological integrity of the forest and the social health of the community. Although businesses succeed or fail for a variety of reasons over which certification by SmartWood may have little effect, certified operations are intended to remain economically viable over the long term. The mandate of SmartWood is to evaluate economic viability from the perspective of ensuring, as much as possible, that sound long-term investments are being made by the forestry operation in terms of management, conservation, and the well-being of local communities. In this sense, it would be a *misinterpretation* to construe SmartWood's responsibility to serve as a financial guarantor of success to investors, shareholders, and/or other parties.

SmartWood certification is focused first and foremost on forests and local communities. The emphasis is how to maximize the value of forest operations in terms of local economies and how to ensure the long-term viability of forest operations. Long-term, economically viable forest management is promoted by the efficient use of multiple products and services and by investing in environmental health and community well-being. To this end, businesses that deal with forest management and the marketing of forest products should encourage the optimal use of a variety of such products.

Utilizing local mills and other local businesses on a regular basis helps to maintain the social and physical infrastructure of the local economy. From an environmental viewpoint, local processing offers the advantage of reducing the energy costs of transporting unfinished products. The bottom line, however, is that certified operations must remain economically viable over the long term.

Economic viability for investments in forestlands cannot be measured solely by quarterly returns. Forestlands do not grow that way; ecosystems do not produce that way; restored land does not recover that way. The economic viability of forestlands is a long-term investment that needs to be measured over generations.

In the 10 years that I (Walter) have been involved with certification, I have visited or assessed over 50 certified forestlands, ranging from 10 to 250,000 acres in the U.S., Canada, and abroad. One common theme among certified forest landowners and managers is their long-term view of their investment. Where they differ from other landowners is their ability to see the return over a much longer time frame, often many lifetimes.

Make no mistake, most of these forestland owners see their forest as a financial investment. Many see the forest as a savings account, where they build their principle (inventory) so that the interest (annual allowable cut) is at its maximum potential yield, and that yield is not only for their personal benefit but also for the benefit of future generations. Of course, as good land stewards, they want to balance the biological and financial yield with other values that the forest provides to the environment and society.

One such landowner bought about 1000 acres of heavily cut-over forestlands in the mid-1950s, and expanded the initial acreage to 1800 acres by 1980. The land had been clear-cut in areas and severely high graded in others. The family worked hard and invested money to restore the land's ecological health and productivity. They did not have their first commercial harvest until 17 years after the initial purchase.

The silviculture was designed to "thin from below," cutting the suppressed and intermediate trees. Because the logs were small and poor in quality, local sawmills were not eager to purchase them. But now, after 40 years of management, their logs are high quality and local mills compete heavily to purchase them. Today, the family enjoys a significant economic return. The local logger who works for them has steady yearly work. And the forest is beginning to have attributes that mimic older forests — high, closed canopies; large diameter stems; large woody debris on the forest floor; and increasing habitat that invites a number of endangered species to call that forest their home.

The property did have residual old-growth trees left by the early loggers because they either posed some logistical logging problem or did not meet the quality standards of the day. The landowners left those trees as legacies of the past forest and have named them after their grandchildren. With this act, the landowner created a form of insurance because it would take a pretty callous member of a future generation to cut down a tree with a family member's name attached to it.

Those old trees, along with a surrounding buffer of younger trees, not only are protected from logging but also are important to fungi, plants, wildlife, insects, and the soil because they are a genetic link to the original forest and a functional link of continuity and soil fertility. The landowners have also left stream side areas untouched, protecting an important fish-bearing stream.

The next generation is actively engaged in the management of the forest. The generation following that is now beginning to get involved and is interested in continuing the tradition.

One may have considered this forestland as an economic failure because it seemed to languish for 17 years without generating any revenue. How many timber companies would consider such a long-term commitment when their investors are clamoring for quarterly returns on their investment?

In the end, however, the investment made by the landowners was a great economic success, one that can continue through generations if wisdom prevails in each generation's forestry practices. Not only is it a success for the family but also for the logger, who is employed every year, and the local mills can count on a certain volume of high quality logs over time.

Keep in mind that the difference between a timber company that operates in the short term vs. a family that operates in the long term is only a matter of choice. And it is these kinds of choices and their social-environmental outcomes that the following criteria are designed to measure.

7.1 — Based on local experience and markets, stumpage or other rents for the use of products derived from the forest are at or above the norm and perceived by landowners as a positive incentive for long-term, biologically sound forest management.

7.2 — The forestry operation is financially viable, and sufficient revenue is generated to cover such costs of management as planning, road maintenance, silvicultural treatments, and monitoring that will maintain and/or enhance the value of the biological services and resources of the forest.

7.3 — Long-term financial planning is evident for the forestry operation.

7.4 — Forestry operations encourage the utilization of frequently occurring, less commonly used species of plants and/or other products for commercial or alternative purposes.

7.5 — Logs and lumber are so handled as to maximize potential usable wood.

7.6 — Managers seek the "highest" and "best" use for trees, both as individuals and as species.

7.8 — Forest management activities strengthen and diversify the local economy.

8 — Product Accountability

A certified forest operation must be able to keep track of the inventory and handling of its product(s) up to the point of sale or transport to other parties outside of the forest.

8.0 — Chain-of-Custody in the Forest

A certified operation must be able to keep track of product inventory and handling up to the point of sale or transport of the product to other parties outside of the forest. All certified products leaving the forest must have "SmartWood certified" or "Forest Stewdardship Council certified" and the joint Forest Management/Chain-of-Custody certification codes printed on them. Good tracking procedures provide customers with a guarantee that SmartWood-certified products originate from well-managed forests.

8.1 — Data are available for the volume and source of the logs that leave the forest (i.e., scaled, inventoried, and measured).

8.2 — Invoices, bills of lading, certificates of origin (e.g., GATT Form A), and other applicable documentation related to shipping or transport specify the certified status of the products.

8.3 — Certified forest products are clearly distinguished from noncertified products through marks or labels, documented separate storage, or accompanying invoices or bills of lading uniquely marked as certified. Such markings must accompany all stages of processing and distribution up to the point of sale or transport either outside of the forest (i.e., up to the "forest gate") or to a third party.

8.4 — All public representation of certified forest products is consistent with the policies of both SmartWood and the Forest Stewdardship Council.

9 — Addendum: Indigenous Peoples within the State of California

The Institute for Sustainable Forestry/SmartWood uses this addendum for the assessment of Indigenous Americans or operations of tribal forestry in California.

9.0 — Indigenous Land Ownerships

9.1 — Tribal peoples, when they so desire, take the lead role in planning forest management operations on land of tribal ownership.

9.2 — Forest management occurs only after securing the informed consent of the tribe(s) or individual tribal member(s) whose forest ownership is being considered for certification.

9.3 — Forest projects operating on land of tribal ownership but under nontribal management must provide documentary evidence of the agreements with the tribe(s) under which nontribal management is entitled to implement operations within the forest.

9.4 — Areas of restricted access on lands of tribal ownership are defined with the full consent and control of the affected tribal members and in accordance with their laws and customs.

9.5 — Forest management shall protect or enhance, either directly or indirectly, the resources and/or the rights of land tenure of the indigenous peoples.

9.6 — The recommendations of tribal representatives are considered during the planning and implementation of management practices to ensure that the land and its uses are protected and/or enhanced by the proposed management.

Phase 6, Group Analysis by the Assessment Team

The final days of each assessment must include closed team meetings for the proposes of

■ Ascertaining what remains to be done and/or learned

Photo 42   Assessors scoring and writing their preliminary report — the Philippines. (Photograph by waitress in hotel.)

- Conducting a group scoring of the guidelines and process of analysis, during which time it is preferable that, at the end of the assessment, scoring of all sections of the guidelines be done as a team (the only way I [Chris] have participated with SmartWood), which means that working together as a team on all aspects of the assessment is strongly encouraged and should be planned for in both time and logistics
- Ascertaining if it is necessary to impose preconditions or conditions on the applicant in order to bring the applicant's operation into compliance with certification standards, where a precondition is a circumstance that must be repaired prior to the possibility of being certified and a condition is a circumstance that must be remedied within a given time frame but which does not prevent certification
- Arriving at preliminary conclusions and recommendations (Photo 42)

Just before leaving the site of the assessment, it is critical for the assessment team to sit down with the applicant and his or her staff to clarify any uncertainties and to discuss the team's preliminary conclusions and potential recommendations. This is what SmartWood refers to as the "reality check."

The reality check is an absolutely essential part of any assessment process because it enables the team to gauge whether its view and potential recommendations are considered valid and feasible by the applicant and staff. All recommendations that are eventually made by the team must be realistic and practical if the conditions of certification are to be successfully implemented. The "reality check" is also undertaken at least once more during the processing of the report in that applicants are given an opportunity to comment on the factual basis, interpretations, and recommendations in the full report once it has been drafted.

Phase 7, Writing the Individual Reports

Following the group analysis by the assessment team and its presentation of preliminary findings in Phase 6, each team member goes home and writes that portion of the total report for which he or she is responsible, be it either a lead or supporting responsibility. It is possible, for example, to have the lead in one area and a supporting role in another, such as the lead in the social arena and a support role in either the environmental or forest management portion of the assessment. Nevertheless, it is absolutely critical that the team leader ensure that each team member is clear about his or her writing responsibilities, including the fact that each person's report is to be completed and submitted to the team leader within 4 weeks after completion of the field work, and the team leader is to have the final report completed 2 weeks later, within 6 weeks of completing the field work. As one can see, good teamwork is crucial to the success of the assessment.

Crafting Individual Reports

For team members, other than the team leader, writing their individual sections usually takes about half the length of time spent in the field (usually 3 to 7 days), whereas for the lead writer, who is normally the team leader, writing and synthesis may take as much time as was spent doing the on-site assessment.

Each team member is responsible for preparing the following:

- Text and analysis corresponding to and arranged by the subject headings in the guidelines
- Text and analysis for topics outside of the framework of the subject headings in the guidelines
- Text and analysis for each criterion in the guideline that is under the purview of the team member; for example, this should describe

the operation's performance vis-à-vis the criterion and be supported by the score assigned to each criterion (see scoring, point two under Phase 6)

- Text and analysis that team members may agree to contribute to sections under another team member's purview
- Summary of findings and main conclusions
- Suggested recommendations (which are nonbinding) for improvement arranged by subject in the guidelines and by priority of application
- Notes and write-ups of interviews conducted by the team member (usually for placement in the appendices)
- Other sections that might be assigned to the team member by the team leader
- Documents to be used as appendices or attachments

The draft sections are submitted to the lead writer for synthesis into a full report. The ultimate responsibility for completing a quality report within the specified time lies with the lead writer and the responsible SmartWood task manager.

Synthesis and Final Report

The lead writer will synthesize the sections from each team member into a full report as specified under the Guidelines for Report Production and Processing of the Forest Stewardship Council and the SmartWood general format for structuring reports, which includes:

- Cover page: SmartWood logo, title, report date, assessment dates, authors
- Table of contents, including list of appendices
- List of team members, with their specialties and affiliations
- Executive summary
- Background, introduction, and assessment methodology
- Full description and analysis of the operation, arranged by the guideline's subject/section headings, while allowing for inclusion of subjects outside of the guidelines
- Full analysis and scoring for each criterion in the guidelines; each criterion should appear within the text for easy interpretation by readers
- Summary of findings and conclusions
- Preconditions, conditions, and recommendations for certification summarized and arranged by priority and time frame for accomplishment and arranged by general subjects/section headings of the guidelines

- Appendices, including: itinerary, list of names and addresses of parties interviewed, interview narratives, maps, photos (always useful), and other relevant documents

The lead writer will synthesize the sections from each team member into a full report that will include:

- Development of a title page, table of contents, appendices, and photographs
- Development of introduction, background, and a section on methodology
- Synthesis and editing of text and analysis to be arranged according to the subject headings of the guidelines, but also allowing the addition of sections that fall outside of the subject headings of the guidelines
- Synthesis and development of a consensus-based full analysis and scoring of the degree to which applicant's operation is in compliance with the guidelines
- Development of a section dealing with the main findings and conclusions
- Development of a section on recommendations
- Development of an executive summary

The lead writer is also responsible for developing a SmartWood Public Summary, based on the executive summary, that can be used as Smart-Wood's public information about the operation. Here it must be understood that the "SmartWood Public Summary" in no way compromises the integrity of the report's confidentiality.

Once completed, the draft report will be circulated for comment by the responsible SmartWood staff to all the team members and such other relevant SmartWood staff and partners as deemed necessary. Following this review, the report will be revised as necessary.

The report will then be circulated by the responsible SmartWood staff to the applicant and his or her staff for comment on the accuracy and feasibility of the recommendations. Then, either simultaneously with the applicant's review or afterward, the SmartWood staff person in charge circulates the report to independent peer reviewers who have had nothing to do with the assessment for a final "reality check," as specified in the policy and procedures.

Phase 8, The Decision to Certify

Once the final report is in the hands of the SmartWood staff, they read it, consider the team's recommendation to certify or not certify the applicant's operation, and make the final decision — after which the applicant is notified of the outcome.

Phase 9, Auditing Procedures to Maintain Forest Certification

Another area of the SmartWood forest certification program is called the "Chain-of-Custody."

Chain-of-Custody in General

Certification is primarily a conservation tool based on a market incentive. Labeling those products that come from sustainably managed forests allows the public to identify and purchase products from companies that reflect the buyer's environmental and social values. These kinds of labeling systems are now in existence on any number of products: recycled products, organically grown food, dolphin-safe tuna, fluorocarbon-free aerosol cans, cosmetics that are free of animal testing, and so on. Increasing numbers of people are choosing products that are environmentally friendly and socially beneficial, as evidenced by the number of new products coming on the market that bear such environmental and/or social claims.

It is imperative, therefore, that certifiers guarantee the public that products bearing the Forest Stewardship Council's label actually come from a forest managed with excellence. To accomplish this task, certifiers have developed a process called "chain-of-custody" certifications. The chain-of-custody certification and auditing process is a way for certification assessors to track certified products through sequential changes of ownership from the landowner, to the mill, to the wholesaler, to the retailer, to the consumer.

Chain-of-custody certification, like forest management certification, is both systems and performance based. Companies that want to produce and/or sell certified forest products must have an inventory tracking system and specific procedures for the storage and handling of the certified wood. These systems and procedures must demonstrate to the assessor that certified products will not be "contaminated" by the co-mingling of non-certified products at any point in the receiving, storage, processing, or retailing process.

In order to maintain the purity of the certified products, the first step is to have an inventory tracking system and storage and handling procedures that verify the origin of the certified products (such as logs, lumber, manufactured parts, or finished products) purchased by the company. Products being received by the company are assigned a Forest Stewardship Council Chain-of-Custody (C-o-C) certificate number. That number must be displayed on all certified product invoices and transportation documents and attached to the product unit. The certificate number, the name of the vendor, the species of tree from which the wood came, and a description of the product (e.g., logs, lumber, chairs, parts of stairs, and so on) must appear on the receiving documents. Those documents must be kept on file along with a summary of the purchases for the assessor to audit.

Second, the storage and handling procedures must be able to verify the movement of certified products through the manufacturing plant, warehouse, or store. Once the certified product has arrived, the company must assign it an inventory tracking code, mark it so that it is visually distinguishable, and store and process it separately from noncertified products. Each time the products change form (logs to lumber, lumber to furniture, and so on) a new inventory number must be assigned to the products, which once again must be visually marked and separately stored.

Third, the destination and volume of certified products sold by a company must be labeled. The label must include the number of the certificate of certification, the Forest Stewardship Council logo, and the company's name. There must also be a sales document on file, as well as a summary of sales that specifies the purchaser of the product, its destination, a description of the product, the kind of wood used, and the volume.

In addition, SmartWood suggests that all company personnel receive training in the procedures for handling and storing certified products, which includes a discussion of their sales and distribution. Sales personnel must be trained in how to represent certified forest products to the public. Written instructions on proper procedures for dealing with certified products must also be provided to the staff.

The process begins with the certifier and assessor when a company sends in an application. The application provides basic information about the company, name and address, type of company (manufacturer, distributor, retailer), products produced and sold, volume sold annually, and so on. This gives the certifier an idea of how the assessment will take place and what questions about the process are important to ask. The assessor then goes to the place of business and performs an on-site assessment of the company's inventory system and procedures for the storage and handling of the materials.

The assessor follows the wood through the entire process. First, wood comes into the company's receiving area. That can be a log yard, lumber yard, warehouse, or other storage area. The assessor interviews the receiving personnel to see if the proper documents come with the shipment, that they are conveyed to the inventory master list, and whether the personnel understand the issues involved with receiving certified products. The assessor inspects the storage of the products to see if the materials are stored separately from noncertified materials and are clearly marked as certified.

Second, the assessor investigates the manufacturing process. The processing line must be cleared of noncertified materials and the certified materials must be separately processed from noncertified materials. The assessor must determine if there are potential points of contamination within the company's facilities. The personnel involved with processing are interviewed again to see if they understand and accept the issues regarding the separation of certified wood from noncertified wood. The manufactured wood must then be reinventoried and receive new inventory documents and be remarked as certified.

Third, the assessor checks with the sales department to determine how they put together the order for sale and shipment. When a shipment is put together, an invoice and shipping document are prepared that have the name of the purchaser, type of product, descriptive detail of the product, kind of wood used, the volume sold, and the price. And internally, the reduction in the inventory of a given product is accounted for and entered into the system.

Finally, the assessor audits the inventory and tracking system in the office. The master inventory printout is reviewed for basic volume of material taken in, which is compared with the volume in material going out. Then the list of vendors, for both products received and products shipped, is reviewed and a random invoice is pulled to see if the information corresponds to the inventory list.

If there are any problems in the system or procedures, as with forest-land certification, preconditions and/or conditions can be initiated with the company.

SmartWood Chain-of-Custody Certification

All SmartWood certifications must, at a minimum, be based on the Smart-Wood Chain-of-Custody Guidelines, which in turn are based on the Forest Stewardship Council's Chain-of-Custody Certification Standards.

The SmartWood Chain-of-Custody Guidelines are used as a public document before, during, and after certification. Chain-of-custody assessors must therefore have extra copies of the guidelines on hand during the assessment process for distribution to interested parties. A representative member or staff person of the SmartWood Network is consulted when questions arise concerning the public nature of the guidelines, communications with the Forest Stewardship Council, or any other step in the certification process.

Companies are certified by SmartWood as either "exclusive" or "nonexclusive." Exclusive companies use wood products in their businesses that have been certified only by a certifier that is accredited through Forest Stewardship Council, whereas nonexclusive companies use both certified and noncertified wood products in their business.

When a company is processing, reselling, manufacturing, or using only certified wood products, the certification process is much simpler and easier. In this case, the assessor's main responsibility is to ensure that the wood being processed as certified comes from a certified source (initially a certified forest). If companies have some products that contain wood from certified sources and different products that contain wood from noncertified sources, they must have a system in place to ensure that the certified wood is tracked and kept pure from contamination with noncertified wood.

A basic tenet of certification endorsed by the Forest Stewardship Council is the notion of reciprocity among certifiers that are accredited through the Forest Stewardship Council. Under this concept, certifiers agree to accept the certifications of all other Forest Stewardship Council-accredited certification programs. For example, a company certified by SmartWood may purchase wood products from a forest or company that has been certified by Scientific Certification Systems and market this product as SmartWood certified.

If, on the other hand, a nonexclusive company is certified through the SmartWood Chain-of-Custody, there is always the potential risk that certified and noncertified wood could become mixed while in the company's facility. It is thus the task of the nonexclusive company to ensure that all steps are taken to minimize the risk of contaminating the flow of certified wood. In turn, it is the job of the assessor to guarantee that the system is both implemented and provides adequate levels of protection against contamination with noncertified wood.

One of the most important sections of any chain-of-custody report is the documentation of the steps taken by a company in handling certified wood, flagging potential steps that may be high risk in nature, and suggesting possible solutions, such as the following:

- Recording the sources, volumes, and descriptions on incoming certified wood
- Recording the internal flow of materials within a production process
- Monitoring processing efficiency and loss of residues in processing, handling, and disposal of residues

- Physical separation of incoming certified wood from incoming noncertified wood
- The attachment of physical means of identification to incoming certified wood
- Physical separation of certified and noncertified intermediate products
- Recording temporal separation of certified wood products from noncertified ones
- Attachment of physical labels to certified end products
- Maintenance of forms to keep tabs on inventory, invoices, bills of lading, customs vouchers, and general accounting ledgers, all with certified labels from the forest and each step to the present

In this way, the company will know what changes to make in order for them to become certified to handle certified wood.

Someone at SmartWood headquarters and/or its network affiliates will answer any question a potential client may have concerning the chain-of-custody certification process and its costs and benefits. When a potential client is ready to move forward with certifying his or her business, he or she must submit an application for chain-of-custody certification. Upon receipt of the application, a staff member from SmartWood headquarters or from a regional SmartWood affiliate assigns a task manager to see the client through the chain-of-custody assessment and certification process. The task manager will be the client's contact person throughout the entire process.

The task manager will prepare an estimated budget for the assessment of the company. This budget will include the cost of the assessor, travel, supplies, and the costs of administrative and technical assistance by SmartWood headquarters and/or a SmartWood Network affiliate. The task manager will submit the budget to the client and, upon approval, prepare a Certification Services Agreement for the client's signature, at which time a request for initial payment to cover the cost of the assessment will be made. With the initial payment in hand, the task manager will then begin the chain-of-custody assessment process.

Beginning the Chain-of-Custody Certification Process

The SmartWood task manager has the responsibility of setting up the chain-of-custody certification. The task manager selects the person who will actually conduct the certification process (i.e., visit the site and write the assessment report). Selection of the assessor is based on many factors, including experience with SmartWood chain-of-custody certification and personal familiarity with the type of operation to be assessed, its geographic location, and prices. If a consultant is selected as an assessor, he or she must fill out and sign a consultant agreement form that outlines the terms of the assessment, payment, confidentiality, and identification of potential conflicts of interest. In case the SmartWood task manager is the actual assessor, he or she must also complete a confidentiality and disclosure form.

Self-Assessment

A client seeking certification for his or her company can conduct an optional self-assessment by examining its internal control systems in relation to the guidelines for SmartWood chain-of-custody certification. When requested, the SmartWood task manager will provide the client with an electronic version of the guidelines on a diskette or via e-mail. The client then proceeds to provide a point-by-point written response to each standard in the guidelines. This information is then returned to the task manager who provides this preliminary information to the assessor.

The self-assessment process is an optional step that has proven valuable in reducing the costs of certification for the client and in ensuring that the company's staff better understands the requirements of and for certification. It has also made the process of pre-assessment preparation much easier. There are cases, however, when it does not make sense for a company to go through a self-assessment or when a company chooses not to do so.

Pre-Assessment Preparation

In order to ensure that the chain-of-custody assessor's time is both efficiently and effectively used during the on-site visit, the task manager works with the company via telephone or written communications to prepare for the assessor's visit by helping the client to:

- Clearly define the scope of the chain-of-custody assessment, i.e., which facilities and products will be covered by the chain-of-custody certification
- Understand instructions concerning the information that must be gathered and made available to the assessor during the on-site visit

It is preferred that all required documents be organized and put in a binder of some kind for review during the assessor's site visit, and it must be stressed to the client that *all of this information is kept strictly confidential*.

The following is provided as a short list of the information that should be made available to the SmartWood assessor in the offices of the company at the start of the assessment process:

- List and location of company staff and a copy of the organizational chart, if it exists
- List and location of the company's facilities, including facilities for processesing, such as sawmills, kilns, warehouses, and so on
- List and location of suppliers
- List and location of contractors used during the production process
- Annual production in volume, species of tree from which wood is derived, and the projected production of each product over the current and next 2 years
- Listing of available confidential or public documentation of the company's operation, including annual reports to the board of directors or shareholders, marketing reports, or documents for public consumption
- A comprehensive list of product buyers

Beyond this, the client must designate a lead contact person within the company whose job it will be to coordinate with the SmartWood assessor. And there is, of course, administrative paperwork.

Administrative Paperwork

It is the job of the SmartWood task manager to ensure that the paperwork for the consulting assessor, such as signed statements of confidentiality and conflict of interest, is completed and on file and that the following internal SmartWood documentation is provided or available to the assessor prior to his or her commencing the process:

- Terms of reference
- Current forms detailing procedures for claiming expenses
- Standard form showing how to prepare an invoice
- Copy of the SmartWood guidelines and a template for writing the assessment report
- Copy of the SmartWood Manual for chain-of-custody assessments and audits
- Copy of the latest description of the SmartWood program
- Copy of the latest SmartWood generic chain-of-custody guidelines or relevant regional guidelines

■ Copy of the latest SmartWood list of certified companies and forest management operations

Next comes the actual on-site assessement.

On-Site Assessment

The on-site chain-of-custody assessment of a company is, without a doubt, the most critical and most complex stage of the chain-of-custody certification process. During a relatively short visit to the company, the assessor must gain a clear understanding of the company's production system and evaluate the integrity of its proposed procedures for meeting the chain-of-custody requirements. For clarity, the following discussion of the on-site assessment is divided into the analytical steps necessary for the successful evaluation of whether a company meets the standards required for chain-of-custody certification.

Step 1 — Company Overview and Scope of Certificate

Overview of company: An assessor must gain a clear understanding of the company seeking certification and develop a precise definition of what products and production facilities will be covered by the chain-of-custody certificate. In broad terms, the assessor must provide a description of the company, its organizational structure, and its overall production, including the use of certified and noncertified products. Information must include the company's location, level of production, source(s) of raw materials, and current markets.

Every effort must be made to get the most current information (e.g., vendors used during the current year, production figures, and so on). If necessary, copies of relevant company documents may be added to the report as appendices.

Defining the scope of the chain-of-custody: The assessor must describe the portion of the company that is being considered for chain-of-custody certification, including a clear description of what products will be produced and which facilities and personnel will be involved in the production. Specifically, information must be included on the certified product to be made, what the source(s) of raw material are, and the proposed markets for the finished products.

Step 2 — Company's Chain-of-Custody Control System

Overview of the company's chain-of-custody control system: The assessor must describe in general terms the elements and strategy upon which the control system is based. Chain-of-custody systems are based on a combination of the following elements:

- Exclusivity (utilization of one certified material in production)
- Physical separation (e.g., procedures are in place for separate processing of certified products; separate facilities are designated for certified-only processing; or areas within a given facility are designated to be used only for certified materials, such as storage areas for logs and/or warehouses)
- Temporal separation (e.g., producing certified products in batches that are separated in time from batches of noncertified products — day shift vs. night shift)
- Labeling or some other method of physically marking certified products
- Selected staff with clear responsibilities for maintaining and monitoring the chain-of-custody system (producer must designate a primary contact person, as well as a contact person at each level of production and in each facility or within each location within a facility, who is responsible for ensuring that the chain-of-custody protocol is being followed)
- System of record keeping that maintains information and documents pertaining to the purchase, processing, and sale of certified products

Certified product flow chart: The certified product flow chart provides a detailed description of the trail taken by the certified products, including steps in handling and system elements for ensuring compliance with chain-of-custody protocol (see Appendix 1 for an example of a certified product flow chart). The flow chart facilitates the identification of potential points at which there is a risk of contamination with noncertified materials and greatly facilitates follow-up auditing. The assessor must develop a flow chart for each path of processing according to the following (note, however, that different products that follow the same steps in manufacturing, handling, and tracking can be combined into a single product flow chart):

- Describe in general terms which specific products are covered by the manufacturing process (e.g., kiln dried dimensional lumber, custom molding, particle board, and so on). This description may include information on the species and grade of wood used.
- Describe the major step from start to finish in the processing and handling of each certified product, which means that the flow chart must begin from the point at which the certified material is received

by the company from a certified source or certified chain-of-custody supplier and end with the point of final sale and/or shipment of the certified product to the purchaser. This information should be presented in a type of product development flow chart that details major steps, procedures, or activities undertaken to protect the purity of the certified materials. The flow chart should also describe how the record keeping and documentation support the chain-of-custody as it is tracked by the company at each step.

SmartWood guidelines and analysis: To complete the examination of the company's control system, the assessor should provide a point-by-point analysis of the operation using the SmartWood chain-of-custody guidelines. For each standard in the guidelines, the assessor should describe the elements of the company's tracking system by addressing each standard, as well as by an assessment of the company's compliance with each standard or a perceived weakness in the company's system.

Step 3 — Analyzing the Risk of Contamination

The assessor should identify and describe the key points at which certified materials may become contaminated with noncertified products. For each of the main points of risk that are identified, the assessor should describe the measures used by the company to control the risk. If the risk control measure is deemed inadequate, the assessor, in consultation with the company, must identify a functionally and cost-effective means of improving the system, some of which follows. Additional details in narrative from may be needed to completely understand either the risk or how to control it (see Appendix 2).

Step 4 — Improving the Control System

If the chain-of-custody analysis identifies a weakness or flaw in the system, the assessor must document the company's proposals to redress such weakness or flaw. The SmartWood program guidelines divide the necessary steps for improvement into three categories, any or all of which an assessor can use in addressing weaknesses or flaws in a company's chain-of-custody system:

- Preconditions, which are requirements that a company must agree to *before* it will be certified by SmartWood
- Conditions, which are requirements that companies must agree with and which must be addressed during the 5- year recertification period

■ Recommendations, which are voluntary actions suggested by the assessor, but which are neither mandated nor required

Having covered the mechanics of the SmartWood certification process, we will now discuss some of the common problems we have encountered over the years. The purpose of the following discussion is to help someone interested in applying for forest certification to understand and thus perhaps to correct potential problems before submitting his or her application.

Chapter 5

Some of the Common Problems Encountered during the Certification Process

We have, over the years, seen some common problems emerge again and again during the assessment process, which we will share with you in the event you are interested in pursuing forest certification. These are problems that people must come to grips with before certification is possible. The notion underpinning this chapter is that someone interested in having his or her forestland certified will have a head start in knowing what to look for and thus the opportunity to correct the problem *before* applying for certification, should one of these problems exist on one's land; this would not only make the certification process easier but also possibly prevent a conditional requirement (which will be explained in detail later) from being applied to one's certification. The problems we will discuss are divided into two general groups: (1) forest landowners and forest managers and (2) chain-of-custody.

Forest Landowners and Forest Managers

There are, as you remember, a number of criteria in the assessment process, not all of which need to be addressed in this discussion. We have, therefore, divided this section into three general categories as follows: (1) forestry, (2) environment, and (3) social and economic.

Forestry

Over the first 10 years of the certification "movement," most of the owners and managers of forestlands who applied for certification and were accepted had already been managing in a manner closely aligned with the stated Principles and Criteria of the Forest Stewardship Council. All these people have the following traits in common: humility, integrity, an open mind, and a willingness to learn — always a willingness to learn.

As we have discussed earlier, change, or more to the point adaptability to change, is a necessity when working with ecosystems to make one's living. Many of the foresters and landowners that I (Walter) have had the honor of meeting during my years as a certification assessor have been adaptable to change because, first and foremost, they have the humility to understand that life is a learning continuum and that conventional wisdom is to be constantly questioned — even their own. I cannot remember hearing a single certified forester or forestland owner say that his or her management is *the* right way to do things. They labor over every decision, and after having made a decision, they watch and evaluate what happens (some call it monitoring).

Many, if not most, of the certified foresters have worked on the same properties for many years, and their "education" comes from watching what happens in the forest following a decision and its attendant action. Although they strive to produce a certain outcome, they have the humility to not presuppose they know exactly what that outcome will be. Their system of adaptive management comes from moving cautiously, paying attention to details, and always believing there is more to learn. And finally, when things do not turn out the way they hoped and perhaps expected, they acknowledge the variance and try to understand what happened and why so they can effect a different and better outcome the next time. They do not, however, deny the unwanted or unexpected outcome. These attributes are essential in the *art* and science of forest management.

Management Planning

Having said all of that, one of the issues we constantly run into in doing certification assessments is that management plans are either nonexistent in written form or very limited in scope. The Forest Stewardship Council is explicit on the point that a forest management plan is an inseparable part of managing a certifiable forest, although flexibility is given for the size and scale of the operation, meaning that the larger the forestry endeavor, the more details are needed in the plan. Of course, the owners and/or managers of certifiable forests nearly always say: "I'd rather be out in the woods doing stuff (monitoring, supervising, fixing roads, planting, etc.) than in an office writing plans," or "It costs too much, and my client(s) does not want to pay for it." Although the points are always well taken, as discussed earlier, a management plan is more than simply a set of directions; it is also a historical record, a blueprint, a plan for monitoring, an educational tool, and in a sense a "last will and testament" for the forest.

What landowner would want to allow his or her land to be handed to another generation without some guidance on how to protect the family heritage, which includes such values as ambiance, healthy populations of wildlife and fish, clean water, clean air, fertile soil, fond memories, and spiritual qualities, in addition to the financial value? Not many. Most certified forestland owners say on the surface that their forestlands are financial investments, which they are, but if one listens closely, these landowners are in some way spiritually tied to their land.

Other landowners have been interested in wildlife, sometimes for hunting, sometimes for viewing, and sometimes just because they are naturally there. One particular landowner burns the forest regularly so that the population of quail remains adequate for hunting. It just so happens that the forest is long-leaf pine, an ecosystem that has been all but extirpated from the South of the U.S. The frequent fires are the natural disturbance regime under which long-leaf pine once thrived. All of these "nontimber" issues must be included in a management plan because they are both a legacy of what a particular family values and important ecological information.

Most management plans that are submitted as part of the certification process are timber oriented: volume, growth and yield data, harvesting plans, road construction plans, and silvicultural prescriptions. Although these are important and necessary elements of a management plan, data on other values need to be documented as well, such as the presence and health of wildlife and fish populations, the amount and distribution of snags and down woody debris, both on land and in streams, indicators of forest health to be used in monitoring, and so on.

How, for instance, would a landowner know if the quail population is increasing or decreasing if he or she did not have a sense of their numbers or the amount and quality of their habitat on the property in question? Such data are really no different than measuring the volume and growth of timber to understand what yields will be over time.

How will future owners know that burning is an important aspect of forest ecology for long-leaf pine, where and when the forestland was last burned, what intensity of burn took place, or what the results of the burn were without a record? The answer is simple: if there is no written management plan and there is a disruption in the continuity of management personnel or ownership that prevents the information from being passed on verbally, no one will know.

All circumstances are different. That is the beauty and excitement of working in forests. There are any number of examples of information that have not been included in management plans. The elements most commonly missing have been: a description of special or critically sensitive and irreplaceable biological areas and rare habitat types (such as stands of old-growth trees); surveys for rare, threatened, endangered, or endemic species of plants and animals and their habitat; issues concerning land-use planning and its landscape-level effects; site quality and the fertility of soils; issues concerning forest composition, structure, and function; issues concerning species diversity and genetic diversity; the presence and conservation of nontimber forest products and other biological resources; an analysis of social and economic impact of the ongoing forest management; and a strategy for considering landscape-level management opportunities.

Including the above elements in a management plan seems a daunting task for many foresters in terms of time and cost. But the issue of cost and time is relatively mute when considering that the biggest cost in doing these in-field data collections is getting the forester in the field in the first place. Most foresters will agree that collecting "timber" data is crucial for the financial part of the business, such as calculating harvest levels, future yields, and so forth; thus, they will be in the forest collecting these data anyway. Adding sets of data to collect is a relatively minor increase in time and money spent if one is prepared to do so before going into the field work, and the long-term benefits of the additional information is at least equal to the information on "timber."

Silviculture

Again, the focus for most foresters and landowners is, for the most part, on timber values. A goal of many owners and managers of certified forestlands is to build standing inventory by growing straight, larger-diameter trees. As one certified forester has said to me (Walter), "Forests are bank accounts; the inventory is the principle, and the growth is the interest. The larger the principle, the more interest is accrued."

Many managers and owners of certified forests in the Pacific Northwest manage their forests primarily by "thinning from below," which means removing the suppressed and intermediate trees, thus leaving the codominate and dominate trees to grow. This is an effective way of building the inventory (principle). This is admirable and viewed positively by

Photo 43 An old-growth Douglas fir that has been struck and wounded by lightning. Because of its wound, the old fir is diseased and dying, which adds critical structural and functional diversity to the forest as the tree progresses through the end of its life cycle, eventually to reinvest its biological capital into the soil. (Photograph by Chris Maser.)

certification assessors when it comes to providing long-term socioeconomic opportunities, but ecologically it must also be viewed with caution.

Although thinning from below mimics natural processes to some degree, injured, diseased, dying, and dead trees are also a necessary part of the forest ecosystem (Photo 43). If, in thinning from below, all such trees are removed, that is tantamount to high-grading the habitat in that it "sanitizes" the stand ecologically with respect to quality habitat for wildlife and the availability of coarse woody debris as a reinvestment of biological capital into the soil to maintain its long-term structure and fertility.

While it has been recognized that straight, larger-diameter trees are a major benefit to the larger landscape by providing habitat that now rarely exists (most landscapes in the world, save a few, have lost their big-diameter trees through conversion to secondary forests, agriculture, or development), the practice of siliviculture should be balanced with the need for structural diversity in the ecological sense. The need for structural diversity differs from place to place, but the constant "cleaning up" of this forest or that should be moderated. It is hard for foresters to allow "growing space" to be taken by perceived "nonproductive" elements of the forest. Nonproductive elements would be "wolf" trees (large, branchy, twisted, rotten, live trees), snags, brush, and noncommercial species.

Certified forestry requires a balance between ecology and production forestry. Where that balance lies depends on the situation that the forest manager encounters. Those who start with heavily logged-over lands will have fewer options for management than those who start with a virgin forest. Another forester told me (Walter) that he incorporates the "theory of scarcity" in his forest management scheme. Basically, his theory is that any element in the forest that is scarce needs to be preserved and nurtured until it reaches some equilibrium with related elements. Although not an end-all of ecological theory, it makes for a rule of thumb that has helped him maintain a structurally diverse forest landscape on the forest that he manages.

Environment

There are, under the general category of "environment," a number of recurring problems based on ignorance of how ecosystems function — *not* malfeasance as some people would have one think. These are the problems we shall discuss, beginning with the level of consciousness.

Level of Consciousness

Perhaps the most ubiquitous problem is the notion that if at first one does not succeed, try, try again — but always in the same way, working from the assumption that if one only persists long enough it will work. There are many instances, however, in which no amount of persistence will do any good because the level of consciousness (thinking) that created a problem in the first place is not the level of consciousness that can fix the problem. To fix the problem, you must elevate your consciousness to a higher level, one that has a better grasp of short-term causes and long-term effects. For example, a timber company (we will call it "Old Company") has an old-growth forest with redwood and Douglas fir in the overstory and tanoak as a minor component of the understory, but tanoak is aggressively adaptable and rapidly becomes the dominant tree when the canopy is opened, either by fire or clear-cut logging. The Old Company, seeking to maximize its monetary gain as fast as possible, clear-cuts almost its entire acreage within 10 years, which clears the way for tanoak to gain prominence.

The shift from conifers to tanoak, while ubiquitous, is gradual and goes unnoticed because the personnel of Old Company are so focused on logging. Then someone surveying the survival rate of the planted conifers suddenly notices the tanoak because it is in competition with the planted trees and is winning, and so Old Company institutes a very costly and relatively futile herbicide campaign with tanoak as the avowed enemy. Finally, having cut 95 percent of the commercially valuable conifers and faced now with a growing sea of tanoak, Old Company sells its holdings to a group of people who want to restore the forest and still make a profit, something that is possible.

The problem is that tanoak is seen as "the enemy" by personnel of the new company (which we will call "New Company") just as it was by personnel of Old Company. This "enemy mentality" needlessly siphons off much needed attention, energy, and economic capital from the real problem of finding alternative methods of logging so as to create the desired forest instead of more tanoak woodlands, or needlessly perpetuating those already in existence. To create the desired forest, however, the employees of New Company must step outside of the mental box of Old Company's forestry thinking and practices, which caused the problem in the first place. They must also understand that the problem did not occur overnight and will not be rectified overnight; in fact, it is always more expensive in time, energy, and money to repair an ecosystem than it is to protect its health and integrity so restoration is not necessary. Be that as it may, whatever New Company decides to do, the effects will take place in the invisible present.

Invisible Present

Another pervasive problem is one's inability to perceive gradual change. Consider that all of us can sense change — the growing light at sunrise, the gathering wind before a thunderstorm, the changing seasons. Some of us can see longer-term events and remember more or less snow last winter compared to the snow of other winters or that spring seemed early in coming this year.

But it is an unusual person who can sense, with any degree of precision, the changes that occur over the decades of his or her life. At this scale of time we tend to think of the world in some sort of "steady state," and we typically underestimate the degree to which change has occurred. We are unable to directly sense slow changes, and we are even more limited in our abilities to interpret their relationships of cause and effect. This being the case, the subtle processes that act quietly and unobtrusively over decades — one timber sale at a time — are hidden and reside in the "invisible present."

The invisible present is the scale of time within which New Company's responsibilities to its stated vision are most evident. Within this scale of time, New Company's properties will change during the owners' lifetimes and the lifetimes of their children and grandchildren. And secreted within this scale of time are cumulative effects, lag periods, and thresholds of the forested ecosystem.

Cumulative Effects, Lag Periods, and Thresholds

A third problem is one's inability to stand at a given point in time and see the small, seemingly innocuous effects of one's actions as they accumulate over time until they suddenly become visible, after which it is too late to retract the action if the outcome is negative with respect to one's intentions. The cumulative effects of New Company's activities will compound in secret to a point that something in the environment shifts dramatically enough for people to see it, just as the hidden, cumulative effects of logging gradually converted redwood/fir forests to tanoak woodlands in the days of Old Company. That shift is defined by a threshold of tolerance in the ecosystem, beyond which the system as people knew it, suddenly visibly becomes something else, usually something that is socially undesirable (such as redwood and Douglas fir being replaced by tanoak). Once the ecosystem shifts, however, the effect of that shift is, more often than not, difficult and costly to reverse, if it is reversible at all (again think of vast areas of already existing, aggressive tanoak).

Once a threshold is crossed, there is no going back to the original condition. It is thus necessary to understand something about the relative fragility of simplified ecosystems (agricultural fields and tree farms) as opposed to the robustness of complex ones (marshes, grasslands, and forests).

Fragile ecosystems can go awry in more ways and can break down more suddenly and with less warning than is likely in robust ecosystems, because fragile systems have a larger number of components with narrow tolerances than do robust ones. As such, the failure of any component can disrupt the system. Therefore, when a pristine ecosystem is altered for human benefit, it is made more fragile, which means that it will require more planning and maintenance to approach the stability of the original system. Thus, while sustainability means maintaining the critical functions performed by the primeval system, or some facsimile thereof, it does not mean restoring or maintaining the primeval condition itself.

To the extent that we alter ecosystems, we make them dependent on our labor to function as we want them to. If we relax our control, they regain their power of self-determined functioning, but usually in ways we do not want.

Let's look at a very simple example that is common to many people: gardening. The more one grooms one's garden to be what one wants, the more specialized the flower beds and vegetable beds become, the more fragile the whole manifests itself as an internally functioning ecosystem. As its fragility increases so does the time and energy one must commit to maintaining the processes one originally disrupted by designing that which was pleasing to one's senses. "But what," you might ask, "was disrupted?"

Consider the act of weeding. A weed is a plant growing where one does not want it to grow. Nevertheless, the weeds in one's garden are an important source of organic material created out of sunlight, carbon dioxide, chemical elements, and water. When they die, this organic material is committed to the soil as dead plants, where it becomes part of the source of energy for the organisms in the soil that are needed to drive and maintain its health. Because the weeds serve a vital function, which one eliminates when one commits the act of "weeding," one must consciously and purposefully put organic material into the soil of one's garden in the form of compost to replace the processes performed naturally by the weeds that one eliminated to maintain one's desired sense of order. Compost, in turn, is made of decomposing plants — even the very weeds that one pulled.

So the more intensely one tries to control one's garden, the more intensely one *must* try to control one's garden if one is to maintain that which one desires; the same is true for a tree farm. This is the self-reinforcing feedback loop that one creates, which is identical to what happened in ancient Greece.

Greece, flourishing under wise agricultural use during the beginning of the Iron Age, had nevertheless greatly altered its landscape, despite its apparently sound agricultural ethic.[38] But all the human-caused changes, including deforestation, do not appear to have caused the collapse of the agricultural system. In fact, it was not only sustainable and was being sustained but also might have continued to the present day if it had not been for the effect of outside influences.

While the Greeks modified their landscape, making it fragile, their agricultural system was sustainable as long as there was a full human population to tend to the terraced fields. The destruction of their agricultural system was not a consequence of the system itself, but rather of

Romans raiding the Greek countryside for slaves, which reduced the population and left the fragile landscape untended to wash into the sea.

Thus, as long as the Greeks maintained adequate cover crops, which were both labor intensive and functioned to hold in place the soil as the forests had once done, their agricultural system was sustainable. But as the activities of Roman slavers continually reduced the Greek population, there came a threshold beyond which this labor-intensive agriculture simply could not be maintained, and the system collapsed.

Prior to the advent of Greek agriculture, the land had been forested for millennia, making sustainability a moot point. Sustainability arose as a problem not because of deforestation, but because of the inability of a society debilitated by slaving to continue performing the function of the forest, namely, soil conservation.

The approaching danger goes undetected, however, until it is too late because ecosystems operate on the basis of lag periods, which simply means there is a lag between the time when the cause of a fundamental change in an ecosystem is introduced (in this case, the mind-set that put Old Company's logging practices in place) and the time when the outcome is visibly apparent. This is somewhat analogous to the incubation period in the human body between contracting a disease and manifesting the symptoms of the disease.

For New Company, the above discussion means that to be sustainable it must recognize and accept that it has relative control over what happens on its land but not absolute control. The more absolutely New Company tries to control its land, the more out of control it will become, just as Old Company did.

New Company, therefore, would be wise to consider that a mountain top with the least human alterations constitutes the most natural end of the continuum, while company headquarters constitutes the most cultural end. Such a continuum can easily be symbolized as follows: N $\longleftrightarrow$ C, where "N" represents the most natural end of the continuum and "C" the most cultural end. Everything in-between, depending on where along the continuum it falls, represents a degree of naturalness and/or a degree of culturalness.

The question for New Company today is where along this continuum must they of necessity maintain a piece of land if the whole of the landscape, in the collective of their individual choices, is to be sustainable — environmentally, economically, and socially. One of the problems of any plan for management of New Company's forest is that, without accounting for cumulative effects, lag periods, and thresholds, there is no way to mitigate actions that would be deleterious to the forest's biological sustainability and thus detrimental over time to the credibility of New

Company's vision and the vision itself, not to mention the company's economic viability in terms of composition, structure, and function.

Composition, Structure, and Function

Here the problem is that people perceive objects by means of their obvious structures or functions but do not understand the role that composition plays in either. Structure is the configuration of elements, parts, or constituents of a thing, be it simple or complex. The structure can be thought of as the organization, arrangement, or makeup of a thing. Function, on the other hand, is what a particular structure either can do or allows to be done to it or with it.

Consider a common object, say, a chair. A chair is a chair because of its structure, which gives it a particular shape. A chair can be characterized as a piece of furniture consisting of a seat, four legs, and a back, and often arms, an object designed to accommodate a sitting person. Because of the seat, we can sit in a chair, and it is the act of sitting, the functional component allowed by the structure, that makes a chair, a chair.

But now we will remove the seat so that the supporting structure on which we sit no longer exists, and now to sit, we must sit on the ground between the legs of the chair. By definition, when we remove a chair's seat, we no longer have a chair, because we have altered the structure and therefore also altered its function. So the structure of an object defines its function, and the function of an object defines its necessary structure, and both add to the ever-widening ripples of diversity. How might the interrelationship of structure and function work in New Company's forestlands?

To maintain ecological functions means that one must maintain the characteristics of the ecosystem in such a way that its processes are sustainable. The characteristics one must be concerned with are (1) composition, (2) structure, (3) function, and (4) Nature's disturbance regimes, which periodically alter an ecosystem's composition, structure, and function.

Nature's disturbance regimes tend to be environmental constraints. True, we can tinker with them, such as the suppression of fire in forests and grasslands, but in the end our tinkering catches up with us and we pay the price.

We can, for example, change the composition of an ecosystem, such as kinds and arrangement of plants within a forest, which means that composition is malleable to human desire and thus negotiable within the context of cause and effect (Photo 44). In this case, composition is the determiner of the structure and function in that composition is the cause of the structure and function rather than the effect of the structure and function.

Photo 44 By clear-cutting the forest, one has grossly altered the composition of plant species that once created the structure and function of the forest. The area in the foreground is now a habitat of grasses and forbs, but is no longer a forest. As the species composition of grasses and forbs changes to include shrubs and, eventually, tree seedlings, the structure and function of the habitat will again change. Each habitat, because of its function that is allowed by its structure that is created by the composition of its plant species, plays host to different animals, and the kinds of animals change as the composition, structure, and function of the habitat changes. (Photograph by Chris Maser.)

Composition determines the structure, and structure determines the function. Thus, by negotiating the composition, one simultaneously negotiates both the structure and function. Once the composition is in place, however, the structure and function are set — unless, of course, the composition is altered, at which time both the structure and function are altered accordingly.

The composition or kinds of plants and their age classes within a plant community create a certain structure that is characteristic of the plant community at any given age. It is the structure of the plant community that in turn creates and maintains certain functions. In addition, it is the composition, structure, and function of a plant community that determines what kinds of animals can live there, how many, and for how long.

If one changes the composition of a forest, one changes the structure, hence the function, and thus one affects the animals. The animals in general are thus ultimately constrained by the composition. Once the composition is ensconced, the structure and its attendant functions operate as a unit in terms of the habitat required for the animals.

People and Nature are continually changing the structure and function of this ecosystem or that ecosystem by manipulating the composition of its plants, which subsequently changes the composition of the animals dependent on the structure and function of the resultant habitat. By altering the composition of plants within an ecosystem, people and Nature alter its structure, which in turn affects how it functions, which in turn determines not only what kinds of animals can live there and how many but also what uses humans can make out of the ecosystem, in this case New Company's property. Therefore, if New Company wants to effect a particular outcome on their land over time, the company must figure out how they want the land to function and then work backward through structure to composition in order to achieve that outcome, which brings us to the notion of ecological redundancy.

Ecological Redundancy

The fifth problem is that people confuse efficiency with effectiveness and therefore seek to reduce and/or eliminate redundancy as both unnecessary and uneconomical. Each ecosystem contains built-in redundancies, which means it contains more than one species that can perform similar functions. Such redundancies give an ecosystem, such as New Company's forestland, the resilience either to resist change and/or to bounce back after disturbance. Redundancy in the biological sense is comprised of the various functions of different species and acts as an environmental insurance policy built into every ecosystem. To maintain this insurance policy, an ecosystem needs diversity of at least three important kinds: biological, genetic, and functional.

New Company would be wise, therefore, to think of each of these kinds of diversity as an individual leg of an old-fashioned, three-legged milking stool because it soon becomes clear that if we lose one leg (one kind of diversity), the stool will fall over. In reality, however, a considerable amount of functional redundancy is built into an ecosystem, such as Old Company's forestland, which means that more than one species (biological diversity passed forward through genetic diversity) can usually perform the same or a very similar function. This results in a stabilizing effect similar to having a six-legged milking stool, but with two legs in each of three locations. Thus, if one leg is removed, it makes no difference which one it is; the stool will remain standing. But if a second leg is removed, the location of the removed leg is crucial, because if it is removed from the same place as the first leg, the stool will fall. If a third leg is removed, the location of the removed leg is even more crucial, because removal has now pushed the system to the limits of its stability, and it is courting ecological collapse in terms of the value Old Company placed on the system in the first place. The removal of one more piece, no matter how well-intentioned, will cause the system to shift dramatically, which in Old Company's case was the reason the land was sold to New Company — so Old Company could avoid having to deal with the tanoak problem they originally caused by eliminating biological, genetic, and functional diversity through their forestry practices.

Now the question for New Company is whether the existing tanoak woodlands are irreversibly ensconced in the landscape despite the amount of money spent trying to reverse the course of management history. If so, then New Company must get outside of the box of mainstream forestry thinking and practices and learn that tanoak can be shaded out by nurturing the best forested areas left on the land rather than cutting them down for an immediate profit. In other words, use the tools Nature has provided rather than throwing more money at the problem, like Old Company did, and try to poison one's way out of the tanoak problem, which only ends up adding poison to the soil and water of New Company's land.

So when one tinkers willy-nilly with an ecosystem's structure to suit one's short-term economic desires, one risks losing species to extinction, either locally or totally, and thus reduces the ecosystem's biodiversity, thus its genetic diversity, and finally its functional diversity. With decreased diversity, one loses choices for safely manipulating one's environment, which, in New Company's case, directly affects New Company's long-term economic viability. The loss of biodiversity may so alter the ecosystem that it can no longer produce that for which it was valued in the first place or potentially could be valued again sometime in the future.

Long-term ecological wholeness and biological richness of the landscape must therefore become a critical measure of economic health because if New Company wants the land to be able to provide for that which the owners desire over time, New Company must do its best to care *first and foremost for the land — beginning immediately*, which means protecting the best forested areas it has left as an ecological blueprint of composition, structure, and function; a refugium of biological and genetic diversity, and a classroom in which to begin learning how to re-forest (not re-tree, but *re-forest*) the many devastated acres they inherited from Old Company. To do this, New Company must understand something about landscape patterns and how its land fits into the whole.

Connectivity and Landscape Patterns

Fragmentation of the landscape is another major problem. The connectivity of landscape patterns is an important consideration because it is not the relationship of numbers, such as acres or boardfeet, that confers stability on ecosystems, but rather the relationship of pattern. Stability flows from the patterns of relationship that have evolved among the various species. A stable, culturally oriented system, even a very diverse one, that fails to support these co-evolved relationships has little chance of being sustainable.

Because ecological sustainability and adaptability depend on the connectivity of the landscape (a seamless forest ecosystem), New Company must ground its properties within Nature's evolved patterns and take advantage of them if it is to have a chance of creating a quality environment in the form of a biologically sustainable forest that is pleasing to the cultural senses of the owners, ecologically adaptable, and thus economically viable.

New Company cannot move away from fragmentation of the habitat on it properties; it can only move *toward* ecological connectivity of habitat. If the owners are to have an adaptable landscape with desirable cultural legacies to pass to their heirs, they must focus on two primary things — and give them primacy: (1) caring for and "managing" for the sustainable connectivity and biological richness in a seamless forest across the landscape, and (2) protecting existing biological, genetic, and functional diversity — including habitats — for the long-term sustainability of the ecological wholeness and the biological richness of the patterns they create across their property.

The spatial patterns on landscapes result from complex interactions among physical, biological, and social forces, in this case meaning "forest management." Most landscapes have also been influenced by the cultural patterns of human use, so the resulting landscape is an ever-changing mosaic of unmanaged and managed patches of habitat, which vary in size, shape, and arrangement.

A disturbance is any relatively discrete event that disrupts the structure of a population and/or community of plants and animals, or disrupts the ecosystem as a whole and thereby changes the availability of resources and/or restructures the physical environment. Cycles of ecological disturbances, ranging from small grass fires to major hurricanes, can be characterized by their distribution in space and the size of disturbance they make, as well as their frequency, duration, intensity, severity, synergism, and predictability.

In the Pacific Northwest, for example, vast areas of unbroken forest (Photo 45) that were at one time in the National Forest System have been

Photo 45 **Unbroken forest like those that once covered the entire forested areas of North America and, more recently, the Pacific Northwest of the U.S. (Photograph by the Oregon Department of Fish and Wildlife.)**

fragmented by clear-cutting (Photo 46) and have been rendered homogeneous by cutting small patches of old-growth timber, by converting these patches into plantations of genetically selected nursery stock, and by leaving small, uncut patches between the clear-cuts. This "staggered-setting system," as it is called, required an extensive network of roads. So before half the land area was cut, almost every water catchment was penetrated by logging roads. And when half the land was cut, the whole of the National Forest System became an all-of-a-piece patchwork quilt with few, if any, forested areas large enough to support those species of birds and mammals that require the interior of the forest as their habitat (Photo 47).

Changing a formerly diverse landscape into a cookie-cutter sameness has profound implications. The spread of such ecological disturbances of Nature as fires, floods, windstorms, and outbreaks of insects, coupled with such disturbances of human society as urbanization and pollution, are important processes in shaping the landscape. The function of those processes is influenced by the diversity of the existing landscape pattern.

Disturbances vary in character and are often controlled by physical features and patterns of vegetation. The variability of each disturbance,

Photo 46 Fragmentation of our forested habitats has been accomplished largely through the practice of clear-cut logging. (Photograph by Chris Maser.)

along with the area's previous history and its particular soil, leads to the existing vegetational mosaic.

The greatest single disturbance to the ecosystem is usually human disruption. These disruptions result most often from our continual and systematic attempts to control the size — minimize the scale — of the various cycles of Nature's disturbance with which the ecosystem has evolved and to which it has become adapted. Among the most obvious is the suppression of fire.

As we humans struggle to minimize the scale of Nature's disturbances in the ecosystem, we alter the system's ability to resist or to cope with the multitude of invisible stresses to which the system adapts through the existence and dynamics of the very cycles of disturbance that we attempt to control. Today's forest fires, for example, are more intense and more extensive than in the past because of the build-up of fuels since the onset of fire suppression. Many forested areas are primed for catastrophic fire. Outbreaks of plant-damaging insects and diseases spread more rapidly over areas of plantation forests or forests that have been stressed through the removal of Nature's own disturbances, to which they are adapted and which control an area's insects and diseases.

The precise mechanisms by which ecosystems cope with stress vary, but one mechanism is tied closely to the genetic selectivity of its species.

Photo 47 When half the land is clear-cut into patches, the whole of the National Forest System will be an all-of-a-piece patchwork quilt with few, if any, forested areas large enough to support those species of birds and mammals that require the interior of the forest as their habitat. Private lands are often more heavily clear-cut than one sees in this photograph, which is public land. (USDA Forest Service photograph by Tom Spies.)

Thus, as an ecosystem changes and is influenced by increasing magnitudes of stresses, the replacement of a stress-sensitive species with a functionally similar but more stress-resistant species preserves the ecosystem's overall productivity. Such replacements of species — redundancy — can result only from evolution within the existing pool of biodiversity. Nature's redundancy must be protected and encouraged.

Human-introduced disturbances, especially fragmentation of habitat, impose stresses with which the ecosystem is ill adapted to cope. Biogeographical studies show that "connectivity" of habitats within the landscape is of prime importance to the persistence of plants and animals in viable numbers in their respective habitats — again, a matter of biodiversity. In this sense the landscape of New Company's holdings must be considered a mosaic of interconnected patches of habitats, which, in the collective, act as corridors or routes of travel between specific patches of suitable habitats, ideally in a forest that is seamless in the management sense.

Whether populations of plants and animals survive in a particular landscape depends on the rate of local extinctions from a patch of habitat

Photo 48 Modifying the connectivity among patches of habitat strongly influences the abundance of species and their patterns of movement. The size, shape, and diversity of patches also influence the patterns of species abundance, and the shape of a patch may determine the species that can use it as habitat. Note, for instance, that the riparian vegetation along the small stream in the upper left quarter of this photograph has been completely removed by clear-cutting. Note also that, while the older clear-cuts are becoming revegetated, they have fragmented the continuity of the habitat because habitat continuity was neither a conscious thought nor an issue when this land was logged. Forest certification is, however, raising the notion of habitat continuity to the fore. (Photograph by Chris Maser.)

and on the rate with which an organism can move among patches of habitat. Those species living in habitats isolated as a result of fragmentation are therefore less likely to persist. Fragmentation of habitat is the most serious threat to biological diversity.

Modifying the connectivity among patches of habitat strongly influences the abundance of species and their patterns of movement (Photo 48). The size, shape, and diversity of patches also influence the patterns of species abundance, and the shape of a patch may determine the species that can use it as habitat. The interaction between the processes of a species' dispersal and the pattern of a landscape determines the temporal dynamics of its populations. Local populations of organisms that can disperse great distances may not be as strongly affected by the spatial arrangement of patches of habitat as are more sedentary species.

The lesson here is that New Company would be wise to have an overall landscape template with which to place its harvest units. Put differently, how harvest units are placed on the landscape in both time and space affects the overall connectivity of the landscape pattern for better or ill. New Company cannot afford to predicate cut units based solely on economics while dealing unconsciously with landscape patterns, as did their predecessor, Old Company. In addition, New Company must account for the scales of diversity.

Scales of Diversity

An issue I (Chris) frequently have to deal with as a consultant is people's lack of understanding biological diversity in time and space across landscapes. Biological diversity in its array of interrelating scales across the time and space of a given landscape is little understood by the general public and is thus a subject of much mistrust when agencies, such as the U.S. Forest Service, attempt to deal with landscape-scale diversity. To help clarify what I mean, two scales of diversity will be discussed, one at the scale of a landscape on public lands and one on small-scale forest holdings on private lands.

Large-Scale Diversity on Public Lands

Some years ago, I was asked to conduct a workshop for the people of the Ouachita National Forest, headquartered in Hot Springs, AR. The problem was the public's concept of an acceptable scale of diversity across the landscape, a concept founded on ignorance of scale and distrust, both of which I found understandable.

For many years, the people had watched, often with a feeling of enraged helplessness, as large timber corporations clear-cut one section of forest after another, converting diverse forests into monocultural plantations of row-cropped trees for the pulp market. Where the people had once seen an acre of forest with a diversity of hardwood trees and shrubs, occasionally with a few conifers mixed in, they were suddenly confronted with row after row of pines. As more and more acres were clear-cut and converted to tree farms for the pulp industry, the people developed a bias against what they perceived as economic simplification of their beloved forest solely for some corporation's short-term monetary gains, which came at the expense of the aesthetic quality of the landscape they felt belonged to all of them.

Although the conversion from forest to single-species tree farm came in two scales, the people were consciously aware of only one, that which encompassed an acre. They saw a diverse forest being converted into a simplified, economic tree farm. What they did not see was the larger picture, which would have been even more disturbing had they recognized it. As industry clear-cut first a few acres and then another few acres, often leaving a few acres of standing forest in between because it did not own the land, industry progressively created a homogeneous landscape as well as homogeneous tree farms, something the U.S. Forest Service had done earlier in the Pacific Northwest.

In the Pacific Northwest, vast areas of unbroken forest that were at one time in our National Forest System have been fragmented by clear-cutting and have been rendered homogeneous by cutting small patches of old-growth timber, by converting these patches into plantations of genetically selected nursery stock, and by leaving small, uncut patches between the clear-cuts. This "staggered-setting system," as it was called, required an extensive network of roads, which meant that before half the land area was cut, almost every water catchment had been penetrated by logging roads. The whole of the National Forest System thus became an all-of-a-piece patchwork quilt with few, if any, forested areas large enough to support those species of birds and mammals that require the interior of the forest as their habitat.

The public who used the Ouachita National Forest, on the other hand, was unknowingly advocating a hidden homogeneity in insisting on all possible diversity on all acres all of the time, theoretically eliminating any disturbance regimes that might create diversity on a larger scale. The public's insistence on small-scale diversity was based, as noted previously, on both a lack of understanding of how the various scales of diversity nest one inside the other and a distrust of the industrial model of "pulp forestry."

Thus, when the folks of the Ouachita National Forest began to restore a single-species pine forest and its simple ground cover of grass along the face of a range of mountains, the public erupted with indignation because they saw it simply as a maneuver to grow an even-aged monoculture of pine trees for the pulp industry. And they had always known they could not trust the Forest Service. That is when I was summoned.

My task was to help all people concerned to understand (1) how exclusively small-scale diversity on all acres all of the time becomes homogeneity of habitat across a landscape; (2) how different scales of diversity not only nest one inside another but also create a collective landscape-scale habitat that is different than the individual habitats created by a single scale of diversity; (3) that the pine/grass community the Forest Service was attempting to restore had indeed existed where the Forest Serivce was attempting to restore it, according to the journals of early settlers, as a fire-induced and -maintained ecosystem in times of pre-European settlement; and (4) the necessity of landscape-scale patterns of diversity if landscapes are to be adaptable to changing conditions and suitable for human habitation over time.

It was critically important for the people to both understand and accept multiple scales of diversity across an array of ecological conditions if the heterogeneity of habitats — and of species — was to be maintained. They needed to understand that diversity is mediated by such events as a falling leaf, a blown-over tree, a fire, a hurricane, a volcano, or an El Niño weather pattern. Each scale of disturbance alters — both destroys and creates — a habitat, or collective of habitats, by renegotiating the composition, structure, and function of plant communities, which in turn allows and creates a time-space array of still different scales, dynamics, and dimensions of diversity that can be used by animals, which in turn can alter plant communities, which in turn become still different habitats, and so on.

Although the workshop took a few days, the people began to change their thinking about the importance and dynamics of large-scale diversity on public lands. Now let's turn our attention to small-scale diversity on private holdings.

Small-Scale Diversity on Private Lands

There are endless possible scenarios of diversity on private lands, but four will suffice in helping to understand the importance of considering multiple scales and dynamics of diversity in relationship to sustainable communities. To keep these examples as comparable as possible, the size of each parcel will be limited to 1,000 acres, and they will be discussed as though you, the reader, owned them.

Example 1

Your forest has a mix of three species of coniferous trees, Douglas fir, western hemlock, and western redcedar, with a few interspersed big-leaf maple and red alder. The forest is relatively uniform in the way it covers the land but is surrounded by a veritable patchwork of industrial clear-cuts, right up to the boundary of your property.

Your thinking, however, stays within the boundary of your property. Thinking your forest lacks diversity, you begin cutting small, dispersed patches of trees throughout the forest to create diversity, but you do so without a clear vision of what you want your forest to look like or why and without any consideration of the conditions of the land surrounding your property, which is all but a sea of clear-cuts. What will your forest look like? Is it what you really wanted? Will you in fact have created more diversity or so fragmented your forest that you have made it even more homogeneous than it was before? How will the pattern you create fit into the surrounding landscape?

In this case, you are creating diversity within the forest canopy for the sake of diversity without taking into account the various scales of diversity. If, for example, small openings are cut in an opportunistic fashion throughout your forest without accounting for diversity at both the scale of your forest and the larger landscape outside of your forest, there will soon be so many little openings and patches of diverse elements within your forest that it will become like alphabet soup out of which no ecological sense can be made. Diversity, to be ecologically viable, must be created in terms of some pattern.

Consider that as the land around your forest is progressively clear-cut, your forest will become a habitat island that will be evermore isolated from other similar habitats. The basic consideration, therefore, must be to make your forest as viable and resilient in its ecological integrity as possible for as long as possible, and that means having a vision within which to create the kind of diversity that is more self-sustaining rather than less self-sustaining, which means understanding the relationship among composition, structure, and function.

To maintain ecological functions means that you must maintain the characteristics of the ecosystem in such a way that its processes are sustainable. The characteristics you must be concerned with are (1) species composition, (2) structure, and (3) function.

You can, for example, change the species composition of your forest, which means the composition is negotiable. In this case, composition is the determiner of the forest's structure and function, in that composition is the cause, rather than the effect, of the structure and function, as previously dicussed.

If, therefore, you want (as part of your vision) a particular animal or group of animals in your forest, you have to work backward by determining what kind of function to create, which means you must know what kind of structure to create, which means you must know what type of species composition of plants (and age classes) is necessary to produce the required habitat(s) for the animal(s) you want. Once the composition is ensconced, the structure and its attendant functions operate as a unit in terms of the habitats required for the animal(s).

People and Nature are continually changing a plant community's structure and function, as well as its attendant animal community, by altering the species composition of the plant community, which in turn affects how it functions. For example, you change the structure of your forest by how and when you cut the trees, which in turn will change the forest's species composition of plants and their age classes, which in turn will change how the forest functions, which in turn will change the kinds and numbers of animals that can live there, albeit much of it is invisible to you belowground or unknown to you in the animal community both above- and belowground. These are the key elements with which you must be concerned, because an effect on one area can, and usually does, simultaneously affect not only your forest but also the entire landscape. Composition, structure, and function work together to create and maintain ecological processes both in time and across space, and it is the health of the processes that in the end creates the health of your forest.

Scale is an often forgotten component of healthy forests and landscapes, however. The treatment of every stand of timber (a designated group of trees) is thus critically important to the health of your forest and the forest landscape as a whole, which can be thought of as a collection of the interrelated stands.

When you focus your whole attention on one stand of trees, you are ignoring the relationship of that particular stand of trees to other stands, to the rest of your forest, and to the surrounding landscape. It is like a jigsaw puzzle where each piece is a stand of trees. The relationship of certain stands of trees in the collective makes a picture of your forest. The relationship of your forest and all surrounding clear-cuts makes a picture of the landscape as a whole.

If one piece is left out of the puzzle, such as the various scales of diversity, the picture on the face of the puzzle is not complete, and you lack the ecological understanding necessary to make your forest as productively sustainable as possible. Your understanding of each stand of trees that you in some way manipulate is therefore critically important in its relationship to the health of your forest as a whole. Therefore, the way each stand is defined and treated is critically important to how your forest, within the surrounding clear-cut landscape, both looks and functions over time.

Example 2

In this example, you own a piece of forestland that, due to past human activities both within and outside of your boundaries, is a fairly homogeneous and predominantly coniferous forest, such as Douglas fir, ponderosa pine, and incense cedar, with a mixture of hardwood trees, such as tanoak and madrone. Being interested in the conditions of your forest prior to European settlement, you check the historical records and find that prior to intensive logging and following World War II, when the planting of conifers replaced the groves of hardwoods, your forest had been predominantly a mixed conifer–hardwood forest with numerous patches of pure hardwoods.

You decide, therefore, to see if you can return your forest to its pre-European settlement condition as you envision it from the literature. How accurate is the literature? Is your parcel of land large enough to sustain itself as a viable ecosystem once you have altered it? How will the structure of your redesigned forest fit as a patch of habitat into the surrounding landscape? Will it be a large enough patch for pre-European settlement animals to live there if they can get to it in the first place? What will happen to the species of animals that live there now? Will you add, lose, or maintain the diversity of species?

It is important for you to understand the island effect that you are creating because restoration of your forest to an earlier condition may severely disrupt the current habitats and existing habitat corridors and replace them with habitats increasingly at odds with the surrounding landscape. Such activity can actually initiate a decline in the viability of an existing local species and/or disrupt existing corridors for uncommon species, thus fragmenting the existing habitat and creating isolated sub-populations that are more prone to local extinctions.

Consider that the spatial patterns you see on the landscape, including those of your own forest, resulted from complex interactions among physical, biological, and social forces. The landscape has been influenced by the cultural patterns of centuries of human use and the resulting landscape is therefore an ever-changing mosaic of patches of habitat, which vary in size, shape, and arrangement. The European-created disruptions of the existing landscape pattern, which included the extensive use of fire by indigenous peoples, began to cause unforeseen changes in the landscape, changes we are now having difficulty dealing with.

Human-introduced disturbances, especially fragmentation of habitat, impose stresses with which an ecosystem may be ill adapted to cope. By restoring the forest on such a small acreage as your land to an earlier condition of habitat, you would be further fragmenting an already fragmented landscape. It is critical to understand this because "connectivity" of habitats with the landscape is of prime importance to the persistence of plants and animals in viable numbers in their respective habitats — again, a matter of both biological and genetic diversity. The landscape must therefore be considered a mosaic of interconnected patches of habitats, which act as corridors or routes of travel between patches of other suitable habitats.

In this sense, your property is probably too small a parcel to restore the "original" condition in such a way that any restored habitat area will be large enough to contain a viable population of most vertebrate species. In addition, restoration of habitats on such a small acreage really means creating new habitats in terms of what now exists and would further fragment the habitat of those species already established in the immediate area because restoration would likely fragment the existing habitat still further in relation to the surrounding habitat. To create a viable culturally oriented forest within your 1,000-acre property, you must not only recognize but also accept, that ecological sustainability and adaptability depend on the connectivity of your forest with the surrounding landscape.

Example 3

In this case, you own a 500-acre tree-farm forest that is primarily Douglas fir with a few scattered western hemlock. The oldest trees are 80 years old. Your management strategy is to cut and remove a number of trees every few years, selecting the biggest and best trees as well as any diseased and dying trees, regardless of age. The number of trees you remove is determined, to some extent at least, by your perceived monetary requirements for the year. Although the trees are felled and removed carefully, so as not to destroy the naturally seeded young trees, you are having a definite effect on the diversity of your forest. What effect are you having on the composition, structure, and function of your tree farm? Are you making it more or less like a forest? What effect are you having on its potential as habitat for wildlife? What effect are you having on the health of the soil?

True, your tree farm superficially looks more like a forest in some respects than do most tree farms in the sense that there is clearly a mixture of age classes among the trees, but you are nonetheless continually simplifying it in terms of species composition, structure, and function through the severity, frequency, and evenness of your manipulations. The multiple entries are constantly altering — and eliminating — habitat for both plants and animals in time and space, such as large dead standing trees and large dead fallen trees. The resulting uniformity excludes some species of organisms that are critical to the long-term ecological health and sustainability of your property as a viable tree farm.

In short, your tree farm is becoming not only an increasingly simplified and homogeneous habitat but also progressively nonsustainable biologically, with less and less resemblance to an ecologically healthy forest. We say this because the simplification of its biological structure through your practice of selective logging amounts to biological "high grading" in which the dominant features of the forest structure and function, as well as the organic material available to the soil, are continually and systematically removed to satisfy immediate profitability at the expense of long-term sustainability.

Here it is important to understand that the less we humans alter an ecosystem, such as your tree farm, to meet our desires, the more the system's functional requirements are met within itself. This in turn makes it easier and less expensive in terms of both time and energy to maintain that system in a relatively steady state because we have maintained more of the diversity of indigenous plants and animals — and therefore biological processes — than we might otherwise have done.

Conversely, the more altered a system is by our human attempt to control it, as in the sense of tree farming, the more that system's functional requirements must be met through human-mediated sources external to itself. This in turn makes it more labor intensive and more expensive to keep that system in a given condition because we have maintained less, often far less, of the diversity of indigenous plants and animals than we would otherwise have done.

Although ecosystems can tolerate cultural alterations, like those you carry out on your tree farm, the ecological functions that are disrupted or removed in the process, often through a loss of species, even locally, must be replaced through human labor if the system is to be sustainable. The more a system, such as your tree farm, is altered and simplified the more fragile it becomes and the more labor intensive becomes its maintenance. When alterations exceed the point at which human labor can maintain the necessary functions, the system collapses. Collapse in this sense means that it becomes something other than that for which it was originally groomed, and in the process it becomes nonproductive of that for which it was altered.

Example 4

In this final example, you are the manager of a recreational forest owned by the city in which you live. You have had the job for 15 years and have watched as urban sprawl has all but surrounded the forest. Now you are faced with not only increasing use of and damage to the forest by people but also an incursion of domestic and feral cats in the forest and holly trees.

Domestic Cats

Domestic cats, both tame and feral (domestic cats gone wild), have invaded the forest; while adding diversity with their presence, they are decimating the small indigenous animals, largely negating the positive effects of sound habitat management. What are your options?

One option, of course, is to do nothing. But that is not viable because the people who spend time in the forest watching birds are beginning to notice an increasingly sharp decline in species, particularly year-round residents, and they demand that something be done about it. Now what are your options?

You could begin by setting up an education program, with the help of the bird-watchers, to inform the townspeople about the effects their cats are having on the local populations of small birds and mammals in the forest. In addition, the bird-watchers and other interested people could gather support and pass a city ordinance that would make it mandatory to have all pet cats neutered. That would substantially control the population. In addition, you could begin a live-trapping and removal program to eliminate as many feral cats as possible roaming the forest.

If that still does not work, the city could pass an ordinance that would make it mandatory to obtain a license for all cats. Anyone's cat caught beyond some distance in the forest would be taken to the local animal shelter and the owner would be notified. An owner would be given two chances to control his or her cat. The third time a cat is caught in the forest, it would be euthanized.

Although this may sound harsh, if the townspeople really value their forest in its entirety, the cats must be controlled as much as possible. The real problem is not the cats but the owners who are irresponsible and allow their cats to stray. Unfortunately, it is usually the cats who pay the price for their owners' thoughtless negligence.

Holly

In addition to cats, holly trees are beginning to show up in the forest in some areas and are taking over the understory at the expense of indigenous plants; in so doing, they are altering the habitat for wildlife. Again, what are your options?

Doing nothing is certainly one of them. But this time, people of the native plant society will not accept that as an option. Now what?

You might find out how holly seeds are dispersed, if you don't already know. You will find that they are dispersed by birds, particularly robins. What options does this information give you? It might stimulate further questions, such as where the robins are getting the holly and how far they fly after eating holly. What ideas would such information give you?

In this case, the solution is fairly simple. Since birds disperse the holly seed during winter, people from the native plant society, with the support of other interested parties, can educate the city council and seek an ordinance to limit and/or remove holly trees within some distance of the forest. This is important because if the source of holly seed is not removed, holly trees in the forest will never be controlled. The notion of control raises the concept of the reinvestment of biological captial.

Reinvestment of Biological Capital

The eighth major problem is that people do not understand the absolute necessity of reinvesting biological capital into the renewable natural resource system from which they make their living. What New Company must keep in mind is that it is continually manipulating the structure and function of its property, as well as its attendant animal community, by altering the species composition of the plant community, which in turn affects how it functions. "Well," the owners might say, "We can't afford to waste merchantable wood. Such waste will affect our investment." New Company has an economic concept of waste, which says in effect that anything not used directly by New Company for its economic benefit is wasted. In a biological sense, however, there is no such thing as waste.

Consider that a tree rotting in a forest is a *re*investment of Nature's biological capital in the long-term maintenance of soil productivity, and hence of the forest itself. Biological capital includes such things as organic material and biological and genetic diversity. And to *re*invest means to invest again.

In a business sense, one makes money (economic capital) and then takes a percentage of those earnings and reinvests them, puts them back as a cost into the maintenance of buildings and equipment, so as to continue making a profit by protecting the integrity of the initial investment over time. In a business, one reinvests economic capital *after* the fact, after the profits have been earned. It is different, however, with biological capital, which is the capital of all renewable natural resources. Biological capital must be reinvested *before* the fact, before the profits are earned.

A forest cannot process economic capital; *biological capital* is required. In a forest, one reinvests biological capital by leaving some proportion of the merchantable trees — both alive and dead — in the forest to rot and recycle themselves and thereby replenish the fabric of the living system.

Such biological reinvestment is *necessary* to maintain the health of the soil, which in large measure equates to the health of the forest. The health of the forest, in turn, equates to the long-term economic health of New Company.

Planting and fertilizing tree seedlings is *no* more of a reinvestment in the soil of the forest than are planting and fertilizing wheat a reinvestment in the soil of a farmer's field. Both are investments in the next crop. As such, they are investments in a potential product, not in the biological sustainability of the living system that produces the products.

As a society, we do not reinvest in maintaining the health of biological processes because we focus on the commercial product. We do not reinvest because we insist that ecological variables, such as the biological health of the soil, are really constant values economically, which can be discounted and therefore need not be considered when it comes to reinvesting economic capital.

But economic capital notwithstanding, all things in Nature are neutral when it comes to any kind of human valuation. Nature has only intrinsic value. So each component of the forest, whether a microscopic bacterium or a towering 800-year-old tree, is thus allowed to develop its prescribed structure, carry out its prescribed function, and interact with other components of the forest through their prescribed interactive, interdependent processes. No component is more or less valuable than any other; each may differ in form, but all are complementary in function.

A forest has three prominent characteristics, as already stated: (1) composition (the variety of plants and animals), (2) structure (how those plants and animals occupy space and time), and (3) function (how those plants and animals interact with one another in creating and maintaining self-reinforcing feedback loops). All of this is incredibly complicated, much like trying to sit on the corner of a water bed without causing ripples.

In a forest one encounters this awesome complexity both aboveground and below. Adding to the apparent complexity aboveground is the internal hidden diversity that comes about when a live old tree eventually becomes injured and/or sickened with disease and begins to die. How a tree dies determines how it decomposes and reinvests its biological capital (organic material, chemical elements, and functional processes) into the soil, and eventually into another forest.

A tree may die standing, only to crumble and fall piecemeal to the floor of the forest over decades. Or, it may fall directly to the floor of the forest as a whole, potentially merchantable tree. Regardless of how it dies, the standing dead tree and fallen tree are only altered states of the live tree, which means that the large, live, old tree must exist before there can be a large, standing, dead tree or large fallen tree.

How a tree dies is important to the health of the forest because its manner of death determines the structural dynamics of the habitat its body provides. Structural dynamics, in turn, determine the biological/chemical diversity hidden within the tree's decomposing body as ecological processes incorporate the old tree into the soil from which the next forest must grow.

What goes on inside the decomposing body of a dying or dead tree is the hidden biological and functional diversity that is totally ignored by economic valuation. That trees become injured and diseased and die is therefore critical to the long-term structural and functional health of the forest, but to an industrial forester such injured and diseased trees are seen only as an economic waste if not cut and converted into money.

The forest is an interconnected, interactive, organic whole defined not by the pieces of its body, but rather by the interdependent functional relationships of those pieces in creating the whole — the intrinsic value of each piece and its complementary function. These processes are all part of Nature's rollover accounting system, which includes such assets as large dead trees, genetic diversity, biological diversity, and functional diversity, all of which count as reinvestments of biological capital in the healthy growing forest.

Intensive, short-term tree farms, which is traditional forestry, disallow reinvestment of biological capital in the soil because such reinvestment has come to be seen erroneously as economic waste. Traditional forestry, which has its root in agriculture, began with the idea that forests, considered only as collections of trees, were perpetual economic producers of wood. With such thinking, it was necessary to convert a tree into some kind of potential economic commodity before it could be assigned a value. In assigning an initial economic value to timber, the health of the soil was ignored. And today, even with our vastly greater scientific knowledge, the health of the soil is not only ignored but also discounted.

Even though protection of soil and its fertility can be justified economically, our human connection with the soil escapes most people. One problem is that traditional linear economics deals with short-term tangible commodities, such as fast-growing trees, rather than with long-term intangible values, such as the future prosperity of our children. But when we recognize that land, labor, and capital are finite and that every ecosystem has a carrying capacity whose needed support in terms of labor and energy depends on its degree of fragility, then we begin to see that the traditional linear economic system is not tenable in the face of biological reality.

Those who analyze the soil by means of traditional linear economic analyses weigh the net worth of protecting the soil only in terms of the expected short-term revenues from future harvests, and they ignore the fact that it is the health of the soil that produces the yields. In short, they see the protection of the soil as a cost with no benefit because the standard method for computing soil expectation values commonly assumes that the productivity of the soil will either remain constant or increase — but never decline.

Given that reasoning, which is both short-sighted and flawed, it is not surprising that those who attempt to manage the land seldom see protection of the soil's productivity as cost effective. But if we could predict the real effects of this economic reasoning on long-term yields, we might have a different view of the invisible costs associated with ignoring the health of the soil.

One of the first steps along the road to protecting the fertility of the soil is to ask how the various ways humans treat an ecosystem affect its long-term productivity, particularly that of the soil itself. Understanding the long-term effects of human activities in turn requires that we know something about what keeps an ecosystem stable and productive, such as habitat diversity and health. With such knowledge, we can turn our often "misplaced genius," as soil scientist David Perry rightly calls it, to the task of maintaining the sustainability and resilience of the soil's fertility. Protecting the soil's fertility is buying an ecological insurance policy for our children.

After all, soil is a bank of elements and water that provides the matrix for the biological processes involved in the cycling of nutrients, which are elements under the right conditions of concentration and availability to plants. In fact, of the 16 chemical elements required for life, plants obtain all but 3, carbon, hydrogen, and oxygen, from the soil. The soil stores these essential nutrients in undecomposed litter and in living tissues and recycles them from one reservoir to another at rates determined by a complex of biological processes and climatic factors.

As soil scientist W. C. Lowdermilk wrote in 1939, "If the soil is destroyed, then our liberty of choice and action is gone, condemning this and future generations to needless privations and dangers." To rectify society's careless actions, Lowdermilk composed what has been called the "Eleventh Commandment," which demands the full and unified attention of New Company if it is to be ecologically sustainable, and hence economically sustainable, to its maximum extent:

> Thou shalt inherit the Holy Earth as a faithful steward, conserving its resources and productivity from generation to generation. Thou shalt safeguard thy fields from soil erosion, thy living waters from drying up, thy forests from desolation, and protect thy hills from overgrazing by thy herds, that thy descendants may have abundance forever. If any shall fail in this stewardship of the land thy fruitful fields shall become sterile stony ground and wasting gullies, and thy descendants shall decrease and live in poverty or perish from off the face of the earth.[39]

With the above in mind, it would behoove New Company to consider how it manages its land if New Company wants to be ecologically healthy and biologically sustainable over time. New Company has to understand, accept, and remember that its forestland must be *biologically* sustainable *before* New Company itself can be *economically* sustainable, which means that New Company must take the scales of diversity into account, even the microscopic scale.

Finally, we come to the last item under Environment, namely, the misunderstood importance of riparian areas and flood plains.

Riparian Areas, Including Floodplains

The ninth and final problem we will discuss is the constant encroachment into riparian areas, including floodplains, in the name of short-term economic profits without understanding the negative long-term ecological consequences of such short-sightedness. Riparian areas can be identified by the presence of vegetation that requires free or unbound water and conditions more moist than normal. These areas may vary considerably in size and the complexity of their vegetative cover because of the many combinations that can be created between the source of water and the physical characteristics of the site. Such characteristics include gradient, aspect of slope, topography, soil, type of stream bottom, quantity and quality of the water, elevation, and the kind of plant community.

Riparian areas have the following things in common: (1) they create well-defined habitats within much drier surrounding areas; (2) they make up a minor portion of the overall area; (3) they are generally more productive than the remainder of the area in terms of the biomass of plants and animals; (4) wildlife use riparian areas disproportionately more than any other type of habitat; and (5) they are a critical source of diversity within an ecosystem.

There are many reasons why riparian areas are so important to wildlife, but not all can be attributed to every area. Each combination of the source of water and the attributes of the site must be considered separately. Some of these reasons are as follows:

1. The presence of water lends importance to the area because habitat for wildlife is composed of food, cover, water, and space. Riparian areas offer one of these critical components, and often all four.
2. The greater availability of water to plants, frequently in combination with deeper soils, increases the production of plant biomass and provides a suitable site for plants that are limited elsewhere by inadequate water. The combination of these factors leads to increased diversity in the species of plants and in the structural and functional diversity of the biotic community.
3. The dramatic contrast between the complex of plants in the riparian area with that of the general surrounding vegetation of the upland forest or grassland adds to the structural diversity of the area. For example, the bank of a stream that is lined with deciduous shrubs and trees provides an edge of stark contrast when surrounded by

coniferous forest or grassland. Moreover, a riparian area dominated by deciduous vegetation provides one kind of habitat in the summer when in full leaf and another type of habitat in the winter following leaf fall.

4. The shape of many riparian areas, particularly the linear nature of streams and rivers, maximizes the development of the edge effect, which is so productive in terms of wildlife.

5. Riparian areas, especially those in coniferous forests, frequently produce more edges within a small area than would otherwise be expected based solely on the structure of the plant communities. In addition, many strata of vegetation are exposed simultaneously in stair-step fashion. This stair-stepping of vegetation of contrasting form (deciduous vs. coniferous, or otherwise evergreen, shrubs, and trees) provides diverse opportunities for feeding and nesting, especially for birds and bats.

6. The microclimate in riparian areas is different from that of the surrounding area because of increased humidity, a higher rate of transpiration (loss of water) from the vegetation, more shade, and increased movement in the air. Some species of animals are particularly attracted to this microclimate.

7. Riparian areas along intermittent and permanent streams and rivers provide routes of migration for wildlife, such as birds, bats, deer, and elk. Deer and elk frequently use these areas as corridors of travel between high-elevation summer ranges and low-elevation winter ranges.

8. Riparian areas, particularly along streams and rivers, may serve as forested connectors between forested habitats or elevational habitats, such as grasslands. Wildlife may use such riparian areas for cover while traveling across otherwise open areas. Some species, especially birds and small mammals, may use such routes in dispersal from their original habitats. This may be caused by the pressures of overpopulation or by shortages of food, cover, or water. Riparian areas provide cover and often provide food and water during such movements.

In addition, riparian areas supply organic material in the form of leaves and twigs that become an important component of the aquatic food chain. Riparian areas also supply shade and large woody debris in the form of fallen trees. While shade keeps the water cool, the large woody debris forms a critical part of the land–water interface, the stability of banks along streams and rivers, and instream habitat for a complex of aquatic plants as well as aquatic invertebrate and vertebrate organisms.[40]

Protecting riparian areas means saving the most diverse, and often the most heavily used, habitat for wildlife. Riparian areas are also an important source of large woody debris for the stream or river whose banks they protect from erosion.[40] Furthermore, riparian areas are periodically flooded in winter, which, along with floodplains, is how a stream or river dissipates part of its energy. It is important that streams and rivers be allowed to dissipate their energy; otherwise, floodwaters could cause considerably more damage than they already do in settled areas.

A floodplain is a plain that borders a stream or river that is subject to flooding. Like riparian areas, floodplains are critical to maintain as open areas because, as the name implies, they frequently flood. These are areas where storm-swollen streams and rivers spread out, decentralizing the velocity of their flow by encountering friction caused by the increased surface area of their temporary bottoms, both of which dissipate much of the floodwater's energy.

It is wise to include floodplains within the matrix of protected areas for several other reasons: (1) they will inevitably flood, which puts any human development at risk, regardless of efforts to steal the floodplain from the stream or river for human use (witness the Mississippi River); (2) they are critical winter habitat for fish;[37] (3) they form important habitat in spring, summer, and autumn for a number of invertebrate and vertebrate wildlife that frequent the water's edge;[40] and (4) they can have important recreational value.

New Company would be wise to learn that if it tries to steal land from a stream or river through the thinking of myopic forestry (as did Old Company), the stream or river will sooner or later reclaim its channel or floodplain, at least temporarily, and potentially at great economic cost to New Company.

It is now time to shift our attention to some of the recurring problems associated with the social and economic aspects of forest certification.

Social and Economic

The social and economic aspects of certification must be divided into private and public forest management. In every corner of the globe, the socioeconomic expectations from private forests are quite different than those from public forests. Although public forestlands, the predominate ownership on a global basis, are required by their very nature to provide more broad-based benefits to society than are private forestlands, private forestlands are not exempt from providing social benefits if the owners want to have them certified under the Principles and Criteria of the Forest Stewardship Council.

Private Forestland

Most owners and managers of forestland bristle at the notion of "social" forestry, e.g., providing the local community some influence over and participation in the forest management, practices, uses, and benefits. Owners of land are particularly sensitive to their "rights of private property" when it comes to how their land is treated; they want the freedom to do as they please. It is not that simple, however, because sustainability requires that land uses by individual property owners have the consensual approval of local communities. If they do not, there will be continual struggle in the form of lawsuits, protests, and/or regulations. Communities around the world have experienced social disruption and violence over land-use practices.

In private forestlands, full control over the management of the land by the public is neither feasible nor appropriate. Be that as it may, many owners and managers of forestlands have told me (Walter) that they do not, for example, want the public to know that they are going to be harvesting because once the chainsaws are fired up and the trucks start rolling, someone is going to notice, and in many such cases, confrontations between landowners, neighbors, and other parties with vested interests are inevitable. On the other hand, owners and managers can do themselves a favor by communicating openly about the operational plans for their forestlands and allow neighbors to get involved in some capacity.

There are many issues, such as economic benefits, noise, dust, truck traffic, safety, aesthetics, ambiance, and changes in micro-climate, that affect neighbors and other parties interested in a given forest management operation. The local community can realize economic benefits from the creation of local jobs through the effects of economic multipliers — forestland owner pays workers, loggers buy from local stores, local stores buy from other stores, and so on. Nothing makes local citizens more upset than to see "foreigners" being hired. Foreigners, in this sense, means people from other areas, such as loggers from California being hired to work in central Oregon; it does *not* mean ethnically or racially "different" local citizens.

There is at least some concept by local people that for forestry and communities to be sustainable they must benefit one another. The more money that is circulated locally from the local forest the more likely it is that the community will be supportive of sustainable forestry and its attendant operations. In the past, we have seen timber companies send profit to distant stockholders (instead of to local citizens in local communities), thereby removing money from the local economy. The local community remains poor, even in resource-rich areas of the world.

Hiring local loggers is generally not a problem for certified operations because owners of forestlands want local loggers who have built a relationship with both the local community and the local forest. They can count on local loggers doing a good job, and most certified companies want their local communities to flourish.

The problem in certification is that private forestland owners do not like being assessed on criteria that rate them on how well they allow the community to interact with their management, how well they treat their employees, and how well they spread the benefits of their forest to the local community. We therefore meet with resistance, even to the point of some landowners choosing not to go through the certification process, even though they are interested, because of the social aspects of the assessment.

In some areas, forestland owners cannot ignore their neighbors because they live in what amounts to a suburban environment. Logging has to be done in these areas with the utmost sensitivity to the neighbors' concerns. Logging must be done during reasonable hours so that noise is not disturbing sleep or weekend relaxation. Restrictions may be imposed because of issues surrounding the flow of traffic as it relates to truck traffic from the logging operation. There also may be issues of safety, particularly when children are walking or being bussed to and from school or curious children and/or adults in the neighborhood wanting to see what is going on.

Again, many private forestland owners believe that their "property" rights allow them to do whatever they want whenever they want to do it, even though common sense would dictate some amount of restraint that would allow the neighbors to at least appreciate, if not support, the forestry operation. One has to remember that "property rights" are granted by the citizens (through laws), because the property "owners" uphold their trusteeship responsibilities. I (Walter) have seen those self-same rights "revoked" by neighbors through lawsuits and direct civil disobedience because the landowner did not make any effort to work with the neighbors or other interested citizens.

Simply affording the neighbors and other interested citizens the common courtesy of seeking their up-front counsel before putting any kind of major forestry operation into motion is enough to forestall immediate confrontations and long, drawn-out conflicts, as well as the tremendous amount of money and time that usually affect both sides. More important than the time and money, however, is the fact that once there has been a breech of trust in a neighborhood, the quality of life diminishes greatly — and irreversibly.

Public Lands

Certification on public forestland is more complex than on private land because there are more parties interested in what is going on and thus more ideas of what should or should not be done on any given piece of ground. In addition, the contention of what should or should not be done on public land is most often ideological and forms the basis for extremely polarized positions that all sides defend with vehemence because each side is absolutely convinced that it is correct and any compromise is a sellout to the opposition. The central problem, however, is that ideological points of view tend to be relatively homogenous and unchanging within a group (the more vehement the group is in defense of its point of view, the more homogenous and entrenched its point of view is), while every piece of ground is different and constantly changing.

Such was the case in south-central Oregon, where a community was struggling to decide how it could best effect a large piece of national forest to benefit its long-term economic stability, which the forest had done as long as there had been enough timber to feed the mills. But now, with only one out of three functional mills left, the local community was faced with an increasingly uncertain future if its citizens could not find a way to replace the lost forestry jobs. To replace these jobs, the community wanted to tackle ecological restoration of the national forest, with forest certification in mind.

Because it was public land, a 3-day conference was held in the community to air the idea, and representatives from the timber company who owned the mill, environmental groups from Oregon and northern California, scientists, people from federal and state agencies, and all interested persons were invited to attend. Before long, it became clear that some people wanted no cutting of timber on public lands, period! Others wanted all old-growth trees to be protected on public lands, industry wanted to continue logging, cattlemen wanted to continue grazing their livestock as in the past, people concerned with endangered trout wanted livestock grazing to cease in riparian zones, others wanted livestock grazing to cease altogether on public lands, and so on. One other thing became eminently clear: the environmental groups were diametrically opposed to forest certification on public lands.

It was finally agreed that a team of people, including a forester, ex-logger, forest ecologist, and stream/range ecologist, would spend time on the ground to assess the current ecological health of the national forest in question. With the field portion of the assessment complete, a report was written and personally presented to the community by the assessment team. The final phase of the assessment was a private meeting of the assessment team and the local committee, whose task it was to seek a solution to the future relationship between the community and the national forest, a solution based on ecological restoration — not on certification. As the meeting progressed, it became necessary that I (Chris) point out to the local committee that the real sticking point was the absence of a clear, shared vision behind which all interested parties could unite and toward which they could all contribute and build. The upshot is that a month or so later I received a telephone call in which I was asked if I would facilitate a process to create a shared vision that would both unite everyone and give everyone something toward which to build.

I agreed, and told the local person with whom I worked that it would require 4 days. I also told him that I wanted some children to participate in the visioning process. The citizen's committee agreed and set a date. In addition to the citizen's committee and some other people from the community, including two or three high school students, there were representatives from a number of environmental groups, the timber and livestock industry, and the local chamber of commerce.

The first day of the visioning process was spent indoors learning about and discussing the ecological functioning of a healthy forest, which included the ecology of wood in soil, the ecology of wood in streams and rivers, the ecological role of fire, the effects of livestock grazing, the kinds and scales of diversity (biological, genetic, and functional from a microbe to a landscape) and how they interact, and how humans alter ecosystems and cause a variety of effects over time. The second and third days were spent in the field examining various types of forests (both healthy and unhealthy), streams in good condition and poor condition, and vantage points from which one could experience an array of landscape patterns at different scales of diversity.

The fourth day was again spent indoors crafting a preamble, vision statement, and goals. There was, as always, much debate concerning the various interests represented in the group. Finally, each person read what he or she had crafted as a proposed vision statement. None worked, however, until a girl who was a junior in high school read her vision statement, which was immediately, intuitively, and unanimously adopted by the group. With an acceptable preamble, vision statement, and goals committed to paper, the group adjourned for 3 months so the preamble, vision statement, and goals could be circulated among a wider representation of the environmental groups, industry, and the local citizens.

When the group met for the final stage of the visioning process, the purpose of which was to revisit the preamble, vision statement, and goals, there were some new people present in the audience. This caused the original debate to be rekindled. After some time, however, it was unanimously decided that the original preamble, vision statement, and goals would stand as written. It was agreed to by the leadership of the national forest that the preamble, vision statement, and goals would work well with national forest management policy and that they were not only acceptable, but something the national forest could and would work with. That done, there was a last bit of work, namely, to draw up a list of tentative objectives and set the date for the next meeting. I then ordered the champagne to be brought out and, while the glasses were being filled, a copy of the preamble, vision statement, and goals was passed around for each person to sign. With that completed and the champagne glasses raised, my work was finished.

As well as in the U.S., management of public forestland in other countries requires a significant effort toward true broad-based participatory involvement of the public. No matter what political system a particular country operates under, SmartWood requires that public lands be managed with the greatest possible benefit both to local communities that depend on a forest's economic opportunities and to the society as a whole.

Public forestlands are managed worldwide by their respective countries under a couple of basic schemes. In some countries, public forestlands are under the direct management of the government in that the government develops management plans, silvicultural practices, areas of protection, and harvesting guidelines through a public involvement and political decision-making process. Once the management plans have been developed, the governmental forest management agency in charge circulates the plans to the public for "input" and "recommendations," after which the agency finalizes the plan and implements it under the authority vested in it by its "public" representatives in the government. Other countries follow a similar pattern, having a management agency vested by the people's representatives to develop the management plan. However, the public's only "input" is through their representative in government.

In the case of the U.S., for example, the governmental forest management agency implements the management plan itself, using its own staff as the management "company," then sells the timber, through a sales contract, to the highest bidder. The successful bidder, in turn, hires the logging contractor to harvest the timber. Other countries, such as Canada, lease forestlands over specified periods of time to forest management operations that actually manage the forestlands, write management plans, harvest plans, and implement silvicultural prescriptions, under governmental guidelines and policy. And then there are countries, such as China, which have government corporations that manage the forestlands under governmental guidelines and policies. Finally, countries like Indonesia have a mix of corporations run by the government, like China, and lease arrangements, like Canada.

Unfortunately, most governments are captured by special interests that exert influence or outright power to get access to forest resources.

Most of the time, these special interests are timber companies; sometimes they are environmental groups. However it turns out, the local communities and greater society lose. This scenario is played out in many communities worldwide.

The extremes of governmental forest policy have manifested themselves in some countries, where there is a high rate of timber theft and the government retaliates by killing the perpetrators. The problem lies in the fact that the governments completely ignore the needs of the people and refuse to allow communities any true participatory management opportunities. In these cases, the local communities are severely depressed economically, and although theft is socially unacceptable, their survival becomes paramount. They thus help themselves to the timber resources.

On the other hand, I (Walter) have witnessed several isolated situations where participatory, community-based management has provided incentives for both society and the local community to protect their native forest. The local community has been given the trusteeship of the forest in exchange for sharing in the economic benefits of that responsibility.

In nearly every case worldwide, high value forest resources and extremely poor communities are synonymous. Until governments move away from special interests to a more participatory and benefit-sharing management scheme, forests will continue to be threatened, both ecologically and socially.

Chain-of-Custody

For certification to remain credible, chain-of-custody tracking of forest products from the forest to the consumer must be as accurate as possible. This means that companies need to prove to the certifying bodies that they can physically separate certified materials from noncertified materials and track the certified materials through the manufacturing or distribution process without the possibility of contamination.

The most common problem for companies is to physically separate certified materials from the noncertified materials. Space is generally limited, which means that companies do not have extra space and/or do not want to spend the money to acquire the necessary extra space. In addition, most companies cannot find enough volume of certified materials to stock their store with 100% certified wood. They must therefore carry separate inventories, one for certified products and one for noncertified products. In many cases, these are the same products, e.g., certified and noncertified two-by-fours or certified and noncertified chairs (with the same species and grade of wood), and so on. The company thus also must put money into buying double inventories.

Certifiers track the product through the distribution network on a monthly or even daily basis. Annual audits are conducted to verify both inputs and outputs. If a company were cheating, selling something labeled certified when it is not, it would be noticed only at the annual audit. However, many consumers (both end users and retailers) are pretty good at questioning the validity of labeled wood. SmartWood gets calls on a regular basis asking for information on companies claiming to be selling certified products. Most of the "rogue" companies, which try to sell noncertified wood as though it were certified, get caught this way.

Conclusions

Most forest managers still focus their attention on the production of timber and its economic value, giving only minor consideration to other forest values, such as wildlife, soil, water, and so on. While companies that are certified fit this generalization, they are doing "light-touch" forestry in that they accept the philosophy of forest certification and are open to incorporating a more holistic approach to forest management based on the best scientific understanding of how forest ecosystems function. Although no forest company measures up completely to the standards of certification, when given conditions for improvement by a SmartWood assessment team or an assessment team from Scientific Certification Systems, or other accredited certifiers, a forest landowner committed to certification is not only willing to meet a threshold of sustainability but also willing to improve over time.

For the future of certification to be strong and for local and regional organizations of certified managers and landowners to eventually become the major force promoting certification itself, landowners and managers must accept their responsibilities as trustees of the land. For this to happen, however, two things must take place: (1) the consuming public must send strong economic signals that they support excellent forestry by preferentially purchasing certified products and (2) local communities and neighbors must support owners of forestlands, foresters, and forestland managers who earn certification for their forests and products. If consumers, local communities, and society in general are willing to support certified landowners, foresters, and managers by purchasing certified products, they can help society take a concrete step toward sustainable community development by helping to safeguard the social-environmental wealth embodied in healthy forests. In so doing, we the adults of the consuming public not only would be passing to our children an increased array of choices but also would be helping to ensure that they have some things of value from which to choose.

Endnotes

1. The following six paragraphs are taken from Frederick J. Deneke. 1998. Forestry: an evolution in consciousness. *Journal of Forestry,* 96(1):56.
2. Chris Maser. 1994. *Sustainable Forestry: Philosophy, Science, and Economics.* St. Lucie Press, Delray Beach, FL. 373 pp. (See pages 77–89.)
3. Greg Brown and Chuck Harris. 1998. Professional foresters and the land ethic, revisited. *Journal of Forestry,* 96(1):4–12; Boris Zeide. 1998. Another look at Leopold's land ethic. *Journal of Forestry,* 96(1):13–19.
4. Karl F. Wenger. 1998. Why manage forests? *Journal of Forestry,* 96(1):1.
5. Jean Mater. 1997. *Reinventing the Forest Industry.* GreenTree Press, Wilsonvill, OR. 263 pp.
6. For an in-depth discussion of the notion of sustainability, see Chris Maser. 1997. *Sustainable Community Development: Principles and Concepts.* St. Lucie Press, Boca Raton, FL. 257 pp.
7. Chris Maser. 1999. *Ecological Diversity in Sustainable Development: The Vital and Forgotten Dimension.* Lewis Publishers, Boca Raton, FL. 401 pp.
8. For a discussion of sustainable forestry per se, see Chris Maser. 1994. *Sustainable Forestry: Philosophy, Science, and Economics.* St. Lucie Press, Boca Raton, FL. 373 pp.
9. P. Bak and K. Chen. 1991. Self-organizing criticality. *Scientific American,* January 1991:46–53; Monica G. Turner. 1989. Landscape ecology: the effect of pattern on process. *Annual Review of Ecological Systems,* 20:171–197.
10. Gifford Pinchot. 1947. *Breaking New Ground.* Harcourt, Brace, New York. 522 pp.
11. The next two paragraphs are based on Wally W. Covington and M.M. Moore. 1991. Changes in Forest Conditions and Multiresource Yields from Ponderosa Pine Forests Since European Settlement, unpublished report submitted to J. Keane, Water Resources Operations, Salt River Project, Phoenix, AZ. 50 pp.
12. D.J. Rapport, H.A. Regier, and T.C. Hutchinson. 1985. Ecosystem behavior under stress. *American Naturalist,* 125:617–640.

13. Larry D. Harris. 1984. *The Fragmented Forest.* University of Chicago Press, Chicago, IL. 211 pp.; Larry D. Harris and Chris Maser. 1984. Animal community characteristics. In: Harris, L.D. 1984. *The Fragmented Forest.* University of Chicago Press, Chicago, IL. pp. 44–68.

14. Michael P. Amaranthus and David A. Perry. 1987. The effect of soil transfers on ectomycorrhizal formation and the survival and growth of conifer seedlings on old, nonforested clear-cuts. *Canadian Journal of Forestry Research,* 17:944–950.

15. F. Thomas Ledig, J. Jesús Vargas-Hernández, and Kurt H. Johnsen. 1998. The conservation of forest genetic resources: case histories from Canada, Mexico, and the United States. *Journal of Forestry,* 96(1):32–41

16. Richard Plochmann. 1989. The forests of Central Europe: a changing view. In: *Oregon's Forestry Outlook: An Uncertain Future.* The 1989 Starker Lectures. Forest Research Laboratory, College of Forestry, Oregon State University, Corvallis. pp. 1–9.

17. For a thorough discussion of the effects of simplifying an ecosystem, see Chris Maser. 1997. *Sustainable Community Development: Principles and Concepts.* St. Lucie Press, Delray Beach, FL. 257 pp.

18. Janet N. Abramovitz. 1997. Learning to value nature's free services. *The Futurist,* 31(4):39–42.

19. The following two discussions of the Gross Domestic Product is based on (1) Timothy R. Campbell. 1998. Sustainable Public Policy: Its Meaning, History, and Application. A paper presented at the annual conference of the Community Development Society in Kansas City, July 19–22. (2) Nicholas Georgescu-Roegen. 1971. *The Entropy Law and the Economic Process.* Harvard University Press, Cambridge, MA and (3) Clifford Cobb, Ted Halstead, and Jonathan Rowe. 1995. If the GDP is up, why is America down? *The Atlantic Monthly,* October:59–60, 62–66.

20. For a thorough discussion of Adam Smith's notion of the "invisible hand," see Chris Maser, Russ Beaton, and Kevin Smith. 1998. *Setting the Stage for Sustainability: A Citizen's Handbook.* Lewis Publishers, Boca Raton, FL. 275 pp.

21. Clyde S. Martin. 1940. Forest resources, cutting practices, and utilization problems in the pine region of the Pacific Northwest. *Journal of Forestry,* 38:681–685.

22. Runsheng Yin, Leon V. Pienaar, and Mary Ellen Aronow. 1998. The productivity and profitability of fiber farming. *Journal of Forestry,* 96(11):13–18.

23. Chris Maser. 1989. *Forest Primeval: The Natural History of an Ancient Forest.* Sierra Club Books, San Francisco, CA. 282 pp.

24. The discussion of carbon storage is based on (1) Mark E. Harmon, William K. Ferrel, and Jerry F. Franklin. 1990. Effects on carbon storage of conversion for old-growth forests to young forests. *Science,* 247:699–702 and (2) J.A. Kershaw, C.D. Oliver, and T.M. Hinckley. 1993. Effect of harvest on old-growth Douglas-fir stands and subsequent management on carbon dioxide levels in the atmosphere. *Journal of Sustainable Forestry,* 1:61–77.

25. Jan Christian Smuts. 1926. *Holism and Evolution.* MacMillan, London. 361 pp.

26. Mark G. Rickenback, David B. Kittredge, Don Dennis, and Tom Stevens. 1998. Ecosystem management: capturing the concept for woodland owners. *Journal of Forestry*, 96(4):18–24.

27. Jeffrey Hayward and Ilan Vertinsky. 1999. High expectations, unexpected benefits: what managers and owners think of certification. *Journal of Forestry*, 97(2):13–17.

28. Sharon T. Friedman. 1999. Forest regeneration practices: how regional certification standards compare. *Journal of Forestry*, 97(2):23–32.

29. The following list is taken from: Catherine M. Mater, V. Alaric Sample, James R. Grace, and Gerald A. Rose. 1999. Third- party, performance-based certification. *Journal of Forestry*, 97(2):6–12.

30. The following list of characteristics is based on Chris Elliott. 1996. Certification as a policy instrument. In: *Certification of Forest Products: Issues and Perspectives*. Virgílio M. Viana, Jamison Ervin, Richard Z. Donovan, Chris Elliott, and Henry Gholz (editors). Island Press, Washington, D.C. pp. 83–91.

31. The discussion of money as a measure of success is based on (1) Peter Lang. 1999. Money as a measure. *Resurgence*, 192:30–31; (2) David Boyle. 1999. The new alchemists. *Resurgence*, 192:32–33.

32. The following discussion of "eco-efficiency" is based on William McDonough and Michael Braungart. 1998. The next Industrial Revolution. *The Atlantic Monthly*, October:82, 83–86, 88–90, 91.

33. The Associated Press. 1999. Washington to launch new master's program. *Albany (OR) Democrat-Herald, Corvallis (OR) Gazette-Times*. January 24.

34. The following discussion of the Genuine Progress Indicator is based on (1) David Orr. 1999. Speed. *Resurgence*, 192:16–20; (2) William McDonough and Michael Braungart. 1998. The next Industrial Revolution. *The Atlantic Monthly*, October:82, 83–86, 88–90, 91; (3) Timothy R. Campbell. 1998. Sustainable Public Policy: Its Meaning, History, and Application. A paper presented at the annual conference of the Community Development Society in Kansas City, July 19–22; (4) Janet N. Abramovitz. 1997. Learning to value nature's free services. *The Futurist*, 31(4):39–42; (5) Gretchen C. Daily, Susan Alexander, Paul R. Ehrlich, Larry Goulder, Jane Lubchenco, and others. 1997. Ecosystem services: benefits supplied to human societies by natural ecosystems. *Issues in Ecology*, 2:1–16.

35. Steve Newman. 1999. Earthweek: a diary of the planet. *Albany (OR) Democrat-Herald, Corvallis (OR) Gazette-Times*. June 6.

36. The following discussion of carbon sequestration is based on (1) Joost Polak. 1999. Storing carbon and cleaning water: how to make profits without cutting trees. *Trendlines*, 1(1):4; and (2) Laurie A. Wayburn. 1999. From theory to practice: increasing carbon stores through forest management. *Pacific Forests*, 2(2):1–2.

37. SmartWood. 1999. SmartWood Forest Assessor Manual. SmartWood Program, Rainforest Alliance, 61 Millet Street, Richmond, VT.

38. Fritz M. Heichelheim. 1956. The effects of classical antiquity on the land. In: W. L. Thomas (Editor). *Man's Role in Changing the Face of the Earth*. University of Chicago Press, Chicago, IL. pp. 165–182.

39. W.C. Lowdermilk. 1975. Conquest of the land through seven thousand years. Agricultural Information Bulletin No. 99, U.S. Department of Agriculture, Soil Conservation Service. U.S. Government Printing Office, Washington, D.C.

40. Chris Maser and James R. Sedell. 1994. *From the Forest to the Sea: The Ecology of Wood in Streams, Rivers, Estuaries, and Oceans.* St. Lucie Press, Delray Beach, FL. 200 pp.

Appendices

Appendix 1   P.T. Asia Mas Construction Material Certified Product Flow Table

Handling Step	Procedure	Certified Product Tracking and Documentation (Exhibit #)
1. Receiving of raw material (logs)	a. Logs are unloaded in log storage area assigned to sawmill #1 b. Logs are checked to incoming Perum Perhutani invoice/shipping manifest c. Data entered into log control record, using PP document number, log size, volume, and number	a. Perum Perhutani shipping document by district (Exhibit A) To be adopted: incoming documentation with logs to be marked by Perum Perhutani as certified; logs to receive company tags and numbers added to control log records; PP included in shipment assigned SW code Internal control for cutting record by log number; pallet control record to follow where material goes to kiln, then to production line #5
2. Sawmill	a. Single log cutting report b. Sawmilling to produce slabs c. Cut to blanks d. Individual pallets tagged; blanks are loaded in pallet	a. Exhibit B for cutting plan b. Exhibit C to be designed to keep control of blanks going into dry kiln and production line #5

Appendix 1 P.T. Asia Mas Construction Material Certified Product Flow Table *(Continued)*

Handling Step	*Procedure*	*Certified Product Tracking and Documentation (Exhibit #)*
3. Kiln drying	a. Load pallets into kiln b. Kiln drying (about 2 weeks, reducing moisture content to 12%) c. Remove pallet from kiln, move into production line #5	a. Same exhibit C
4. Furniture production — component manufacture	a. Distribute the blanks according to process needed b. Make furniture components c. Component order generated by product, record number of components d. Manufactured components inspected, color matching; rejects for resawing, extras into stock e. Daily machining report of components made	
5. Furniture assembly; packing and shipping	a. Furniture assembling b. Final sanding, inspection c. Packing for export d. Load in container	

Appendix 2   Analysis of Risk of Product Contamination

Point of Possible Contamination	Description of Risk	Risk control measure
1. Log yard	Mixing of certified and noncertified logs (risk: low)	Physical checking of incoming logs as to their Perum Perhutani district of origin; physical separation of logs from different districts in the log yard; color end of incoming certified logs
2. Sawmill	Mixing of certified with noncertified lumber (risk: medium)	Separate (separation wall) the log storage area between the two existing sawmills; paint logs ends of certified material to clearly distinguish them from noncertified material
3. Kiln	Mixing of certified with noncertified lumber (risk: low)	a. Load certified and noncertified lumber on separate kiln chamber and/or by pallets b. Mark the certified material to differentiate it from noncertified
4. Furniture factory	Mixing of certified with noncertified components (risk: medium)	a. Keep certified material in separate production line (#5) from incoming to outgoing b. Batch process certified material and mark outgoing components in some way
5. Packing and shipping	Mixing of certified and noncertified finished products (risk: low)	a. Mark with sticker boxes containing certified furniture b. Tag individually packed furniture to indicate certified status
6. Record keeping	Mixing of certified and noncertified documents (risk: low)	a. Mark certified documents that relate to certified materials and products